# Complex Variables
## *Practical Applications*

by D. James Benton

Copyright © 2019 by D. James Benton, all rights reserved.

# Foreword

Complex variables are not simply a mathematical curiosity or academic exercise. They are marvelously useful tools that can open up new solutions to a wide variety of practical problems. After all, mathematics is far more like a language of logic than a collection of facts and figures. Complex variables allow us to handle many two-dimensional problems as if they were merely one-dimensional. They also help us to see that problems in differing fields of applied science are similar and may even unfold with the same solution techniques. Lessons learned and solutions found in one field leading to applications in another is like mathematical recycling! Join me on a tour of this fascinating field of applied mathematics.

*All of the examples contained in this book,*
*(as well as a lot of free programs) are available at...*
<u>https://www.dudleybenton.altervista.org/software/index.html</u>

## Programming

Many of the examples in this book are implemented in the C programming language. Others are implemented in FORTRAN, as this is the only language to have complex variables as a native type. For decades all complex variable applications were written in FORTRAN, which is why there are so many more examples available on the Web.

The Microsoft FORTRAN compilers were positively dreadful, full of bugs and slower than a herd of snails stampeding up the side of a salt dome in a hailstorm. Digital Equipment Corporation took over M$ PowerStation in 1997 and made huge improvements, but the company withered for lack of vision. Compaq took up the torch in 2001, but also failed at the corporate level by not entering the 21st century.

In latter part of this time frame, I was part of a Team performing a study for the U.S. Department of Energy, the Environmental Protection Agency, and the Army Corps of Engineers, evaluating the optimization and parallelization capabilities of several FORTRAN compilers, including Intel's. The study failed to yield any useful results, besides the consistent lack of these compilers to deliver promised performance.

The High Performance Computing Center at the Oak Ridge National Laboratory has done a lot of work along the same vein as this joint DOE, EPA, and ACE project. One of the Team members was employed at that facility. The HPC scientists work very hard to find applications that benefit from specialized compilers and massively parallel machines. Of course, this is the very opposite of demonstrating that useful intensive applications run on extraordinarily expensive and specialized hardware can be cost-effective. This exercise reminds me of designing exotic bolts to fit a wrench you already have.

**Natural Draft Cooling Tower - see analytical solution on page iv**

## Table of Contents

|  | page |
|---|---|
| Foreword | i |
| Programming | i |
| Chapter 1. Introduction | 1 |
| Chapter 2. Riemann Surfaces | 5 |
| Chapter 3. 2D Potentials | 10 |
| Chapter 4. 3D Potentials | 26 |
| Chapter 5. Conformal Mapping | 29 |
| Chapter 6. Polynomial Roots | 43 |
| Chapter 7. A/C Circuits | 53 |
| Chapter 8. Simultaneous Equations | 60 |
| Chapter 9. Ordinary Differential Equations | 63 |
| Appendix A: C++ Implementation | 69 |
| Appendix B. Common Complex Functions | 71 |
| Appendix C: Lambert's W | 73 |
| Appendix D. Error Function | 74 |
| Appendix E: Airy Function | 77 |
| Appendix F. Bessel Functions | 83 |
| Appendix G. Weierstrass Elliptic Functions | 91 |
| Appendix H. Crazy Circular Plots | 93 |

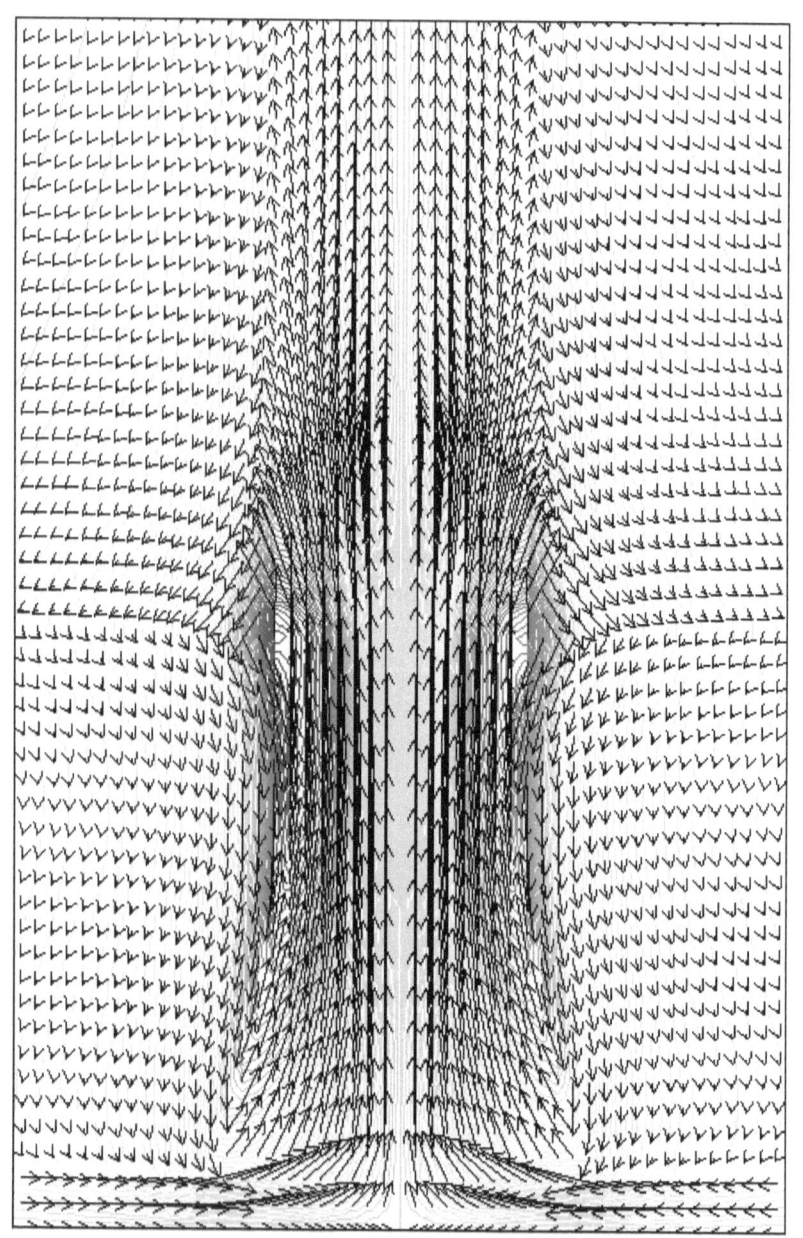

analytical solution for inviscid flow through a natural draft cooling tower
source code is examples\cooling tower\cooling_tower.c

# Chapter 1. Introduction

We must begin by defining imaginary numbers. These have their basis in the square-root of minus one ($\sqrt{-1}$). The designation *imaginary* is somewhat of a misnomer. While we can't fill a pen with $\sqrt{-36}$ sheep in the same way we might with a ram and five ewes, this isn't as nonsensical as it might seem. We readily discuss bank over-drafts, in spite of the fact that we know there aren't negative six hundred dollars (-$600) in our account. It's simply a number. The negative means we owe the bank $600 instead of the bank holding that amount in reserve to cover our next draw.

The square root of minus one is that number, which when squared, is equal to minus one. A similar confusion arises in computer science with minus one. In 8-bit binary this is 11111111, not 10000000. It's the number that becomes zero when we add 1 to it. This works with the first; whereas, the second becomes 10000001, which is 129 in decimal notation. In 8-bit binary, -1 and 255 are the same. In fact, the processor treats them the same, only the state flags change (sign, carry, overflow, and equal, which are true or false, 0 or 1). As we might suspect, there is no binary integer when squared is equal to -1.[1]

In one sense, all numbers are imaginary, because we *imagine* them and they are abstract concepts. If they didn't have any correspondence to things in our experience, this would be a useless academic exercise. Thankfully, this is not the case and imaginary numbers are useful, especially when grouped with a real (i.e., ordinary) number to form a pair we call *complex* numbers.

At the very least, complex numbers are a special type of two-dimensional vectors. In two spatial dimensions, the respective axes are always independent. Complex numbers are different in that even basic operations can result in transfer of information from one component to the other. There are only a few such operations requiring definition. The rest can be derived from these few.

We will use the symbol $i$ to denote $\sqrt{-1}$. Complex numbers consist of a real and an imaginary part, most often written:

$$z = x + iy \qquad (1.1)$$

Addition and subtraction are implemented the same:

$$(a+ib)+(c+id) = (a+c)+i(b+d) \qquad (1.2)$$

Multiplication is only slightly more involved:

$$(a+ib)(c+id) = (ac-bd)+i(ad+bc) \qquad (1.3)$$

Division is a little more involved, but not too difficult:

---

[1] $00001111^2 = 11100001$ (0x0F$^2$=0xE1) and $00010000^2 = 100000000$ (0x10$^2$=0x100) All other numbers are either smaller or larger so that none results in 11111111 (0xFF) when squared.

$$\frac{(a+ib)}{(c+id)} = \frac{(ac+bd)}{(c^2+d^2)} + i\frac{(bc-ad)}{(c^2+d^2)} \tag{1.4}$$

### Euler's Formula

This formula, named after Swiss mathematician Leonhard Euler[2], is both elegant and useful.

$$e^{ix} = \cos x + i\sin x \tag{1.5}$$

While there are various proofs, the simplest one that jumps off the page and slaps us in the face is to compare the three Taylor series:

$$e^x = 1 + x + \frac{x^2}{2!} + \frac{x^3}{3!} + \frac{x^4}{4!} + \frac{x^5}{5!} + \ldots$$
$$\sin x = x - \frac{x^3}{3!} + \frac{x^5}{5!} + \ldots \tag{1.6}$$
$$\cos x = 1 - \frac{x^2}{2!} + \frac{x^4}{4!} + \ldots$$

The even terms contain $i^2=-1$ and so the second and third combine to form the first. Equation 1.5 is the basis for deriving all other exponential relationships. We can also write Equation 1 in terms of radius, $r$, and angle $\theta$.

$$z = x + iy = re^{\theta}$$
$$r = \sqrt{x^2 + y^2} \tag{1.7}$$
$$\theta = \tan^{-1}\left(\frac{y}{x}\right)$$

### The Complex Plane

The complex plane is usually represented by a two-dimensional Cartesian grid, having $x$ as the real axis and $y$ as the imaginary.

---

[2] Leonard Euler (1707-1783) Swiss mathematician, physicist, astronomer, logician, and engineer greatly advanced science and mathematics.

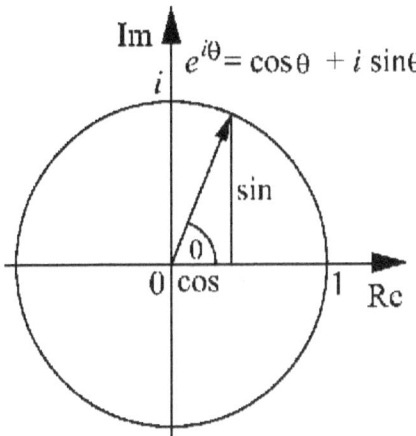

Exponentiation is most easily understood from the complex plane. Equation 1.3 can be used to find the square:

$$(a+ib)^2 = (a^2 - b^2) + i(2ab) \tag{1.8}$$

Equation 1.7 is used to find the polar equivalent:

$$(a+ib)^2 = \left(re^{i\theta}\right)^2 = r^2 e^{i2\theta} \tag{1.9}$$

Substitution easily verifies this result. The square root is also apparent:

$$\sqrt{a+ib} = \left(re^{i\theta}\right)^{\frac{1}{2}} = \sqrt{r}\, e^{\frac{i\theta}{2}} \tag{1.10}$$

Exponentiation has already been utilized in the preceding calculations. The explicit relationship is:

$$e^{(a+ib)} = e^r (\cos\theta + i\sin\theta) \tag{1.11}$$

The natural and common log follow logically:

$$\ln(a+ib) = \ln r (\cos\theta + i\sin\theta) \tag{1.12}$$

$$\log(a+ib) = \log r (\cos\theta + i\sin\theta) \tag{1.13}$$

These operations are most efficiently implemented in C++. The code may be found in the online archive and is described in Appendix A. You will find a program illustrating these in the online archive examples\basics. The output is:

```
examples\basics>basics
z=2+3i=3.60555*exp(0.982794i)
z^2=-5+12i=13*exp(1.96559i)
z^3=-46+9i=46.8722*exp(2.94838i)
z^2.5=-19.1225+15.6099i=24.6848*exp(2.45698i)
sqrt(z)=1.67415+0.895977i=1.89883*exp(0.491397i)
```

```
(2+3i)^(1.5+2.5i)=-0.0187775-0.586381i=0.586682
    *exp(-1.60281i)
exp(2+3i)=-7.31511+1.04274i=7.38906*exp(3i)
ln(1.5+2.5i)=1.07003+1.03038i=1.48548*exp(0.76652i)
log(1.5+2.5i)=0.464709+1.03038i=1.13032*exp(1.1471i)
```

Several transcendental and special functions are covered in Appendix B.

## Chapter 2. Riemann Surfaces

We will now see how several simple complex functions behave by generating surface plots, where $z=f(x,iy)$, taking either the real or imaginary component or some combination of the two. These are called Riemann[3] surfaces and they are quite interesting. This first figure is the real part of $ln(z)$.

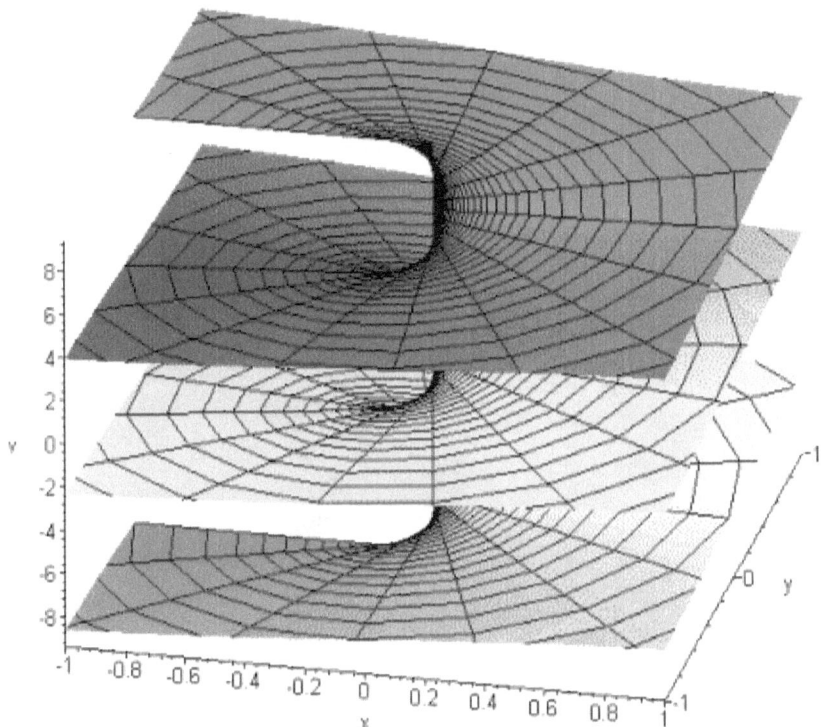

The code to produce this and the following plots can be found in the online archive in folder examples\Riemann surfaces\rsurf.cpp. The output can take several forms, the simplest and most generic is a .3DV file:

```
2500 nodes
-0.00225847,-0.00102154,-9
-0.00174091,-0.00176449,-8.63265
-0.000991064,-0.002272,-8.26531
etc...
2401 elements
1 2 52 51 RGB=0x0010FC
2 3 53 52 RGB=0x0028FC
```

---

[3] Georg Friedrich Bernhard Riemann (1826-1866) German mathematician who made significant contributions in several of areas of theoretical mathematics.

```
3 4 54 53 RGB=0x0038FC
etc...
```

This little program (rsurf.cpp) will output 3D surfaces or closed bodies, depending on which lines of code we select with conditional compilation statements (i.e., #ifdef...). The gamma function is described and the surface displayed in Appendix B. The Lambert function is given by:

$$z = (x+iy)e^{(x+iy)} \qquad (2.1)$$

The Riemann surface of the Lambert function (see Appendix B) with the third dimension being the imaginary component is:

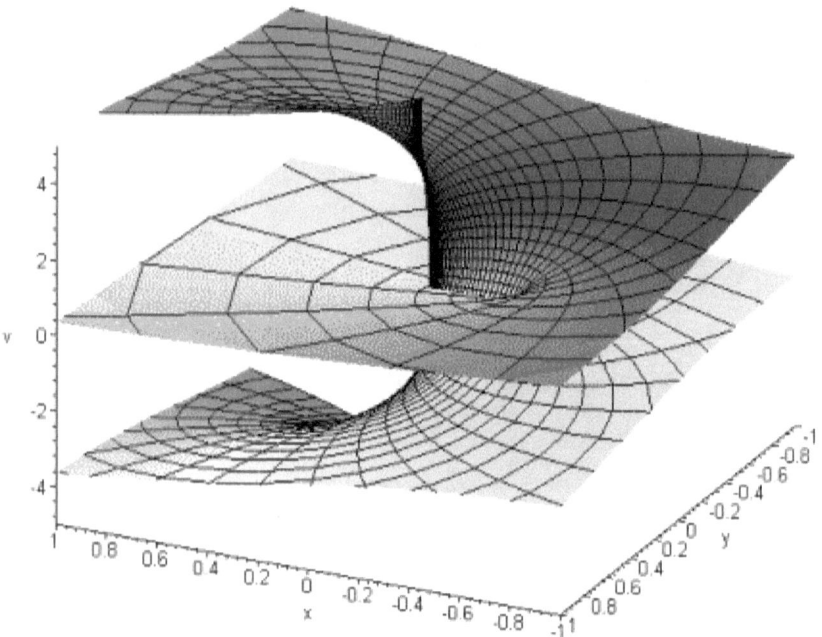

The Dilbert function (not sure how this relates to Scott Adams' wonderful comic) is similar and given by:

$$z = (x+iy)e^{(x+iy)^2} \qquad (2.2)$$

The Riemann surface of the Dilbert function with the third dimension being the imaginary component is:

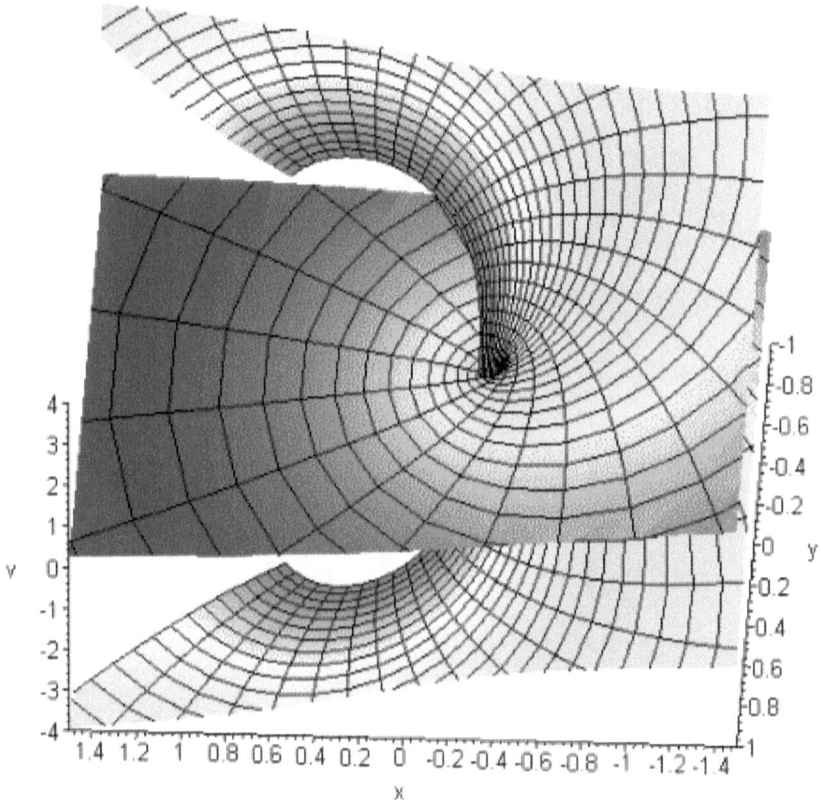

The complex *sin(x+iy)* is shown in this next figure:

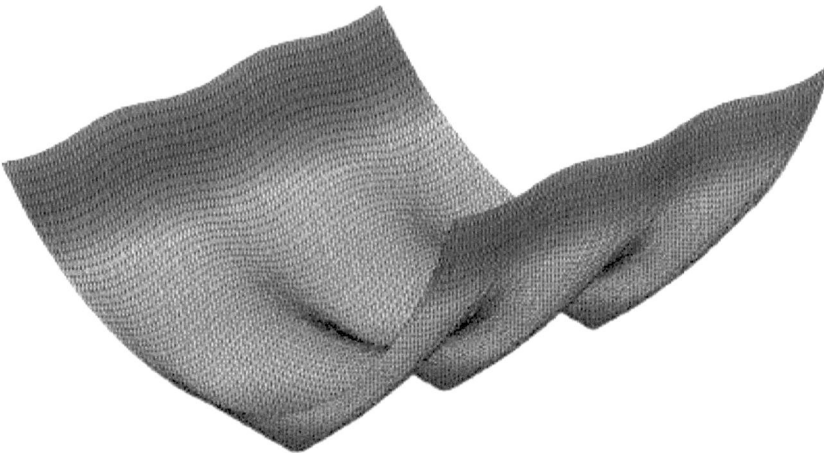

The real part of ln(cos(z)) is shown in this next figure:

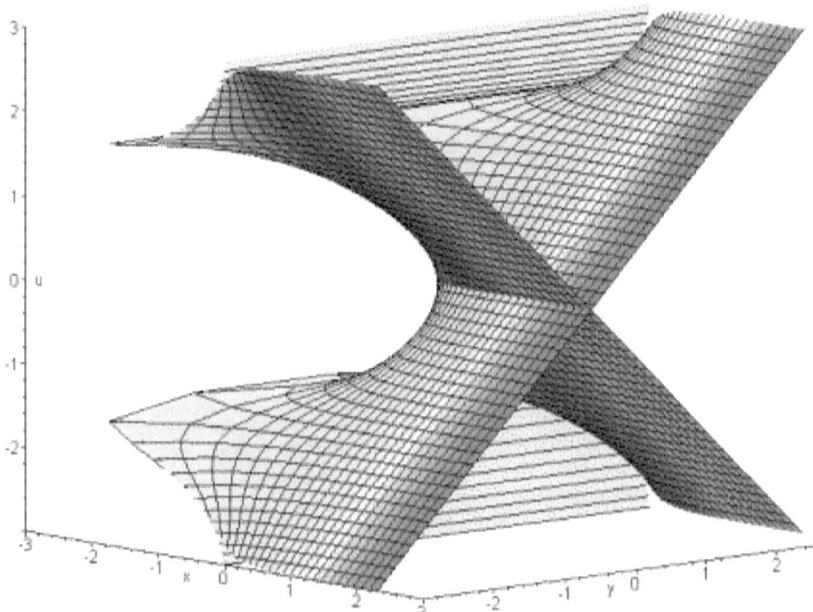

The complex part of ln(cos(z)) is shown in this next figure:

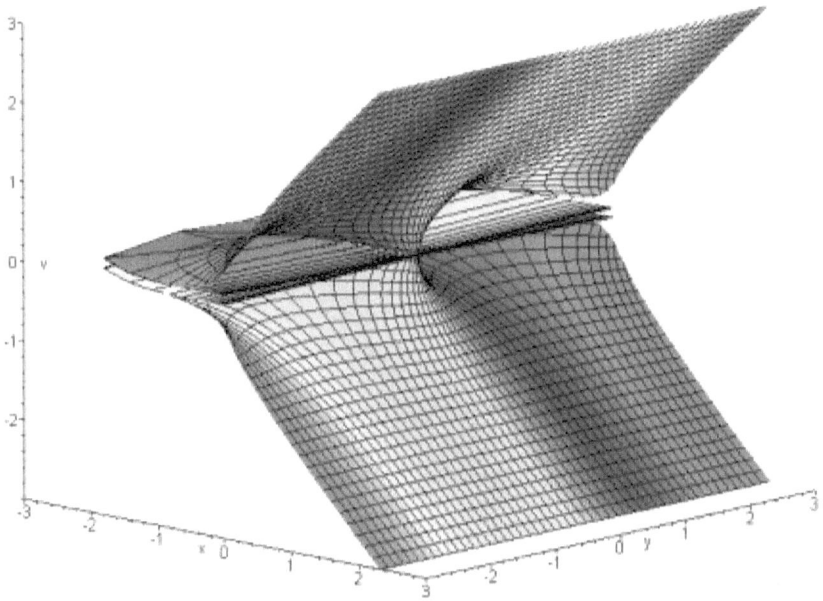

## Chapter 3. 2D Potentials

One of the most fascinating applications of complex variables is linear field theory. Potential is most often given the symbol φ (Greek phi). Such fields satisfy Laplace's equation.[4] In two dimensions this can be written in Cartesian and cylindrical coordinates:

$$\frac{\partial^2 \phi}{\partial x^2} + \frac{\partial^2 \phi}{\partial y^2} = \frac{\partial \phi}{r \partial r} + \frac{\partial^2 \phi}{\partial r^2} + \frac{\partial^2 \phi}{\partial \theta^2} = 0 \qquad (3.1)$$

Applications include electric, magnetic, and gravitational fields, heat conduction, mass diffusion, and inviscid fluid flow. Inviscid (i.e., non-viscous) fluid flow means no viscosity.[5,6,7] We will cover three-dimensional applications in the next chapter. The most familiar example of a potential field is:

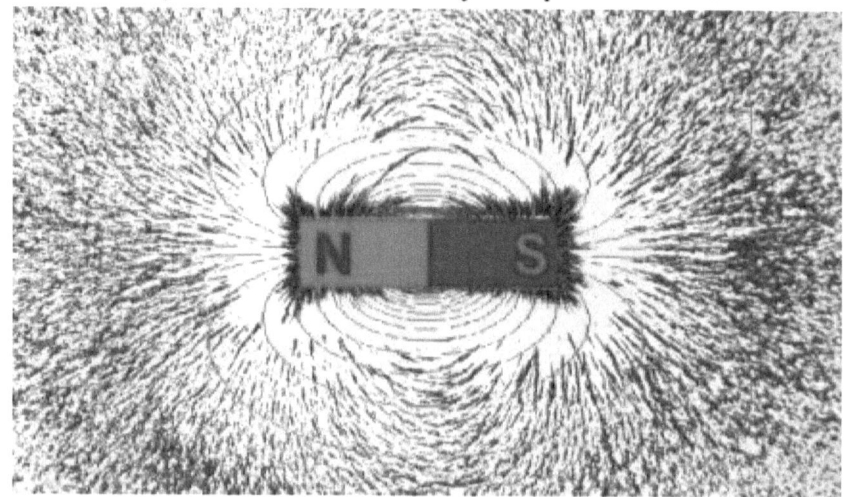

### Streamlines and the Stream Function

Streamlines are the paths that a particle would take if traveling through the field. The stream function is most often given the symbol ψ (Greek psi). The

---

[4] Pierre-Simon Laplace (1749-1827) was a French mathematician, physicist, and astronomer.
[5] This is also called infinite Reynolds number flow, as this dimensionless number is equal to length times velocity times density divided by viscosity. When the viscosity is zero, the Reynolds number is infinite.
[6] George Stokes (1819-1903) was an Irish physicist and mathematician who introduced the Reynolds number, among many other things.
[7] Osborne Reynolds (1842-1912) was an Irish innovator who greatly advanced our understanding of fluid flow.

stream function is perpendicular to the potential function. In Cartesian coordinates this relationship can be expressed:

$$\frac{\partial \psi}{\partial x} = -\frac{\partial \phi}{\partial y}$$
$$\frac{\partial \psi}{\partial y} = \frac{\partial \phi}{\partial x} \tag{3.2}$$

Fluid flow must be two-dimensional and also constant density as well as incompressible in order to be represented by a velocity potential. If the flow is also irrotational, it can be represented by a stream function. The corresponding stream function for the preceding magnet is:

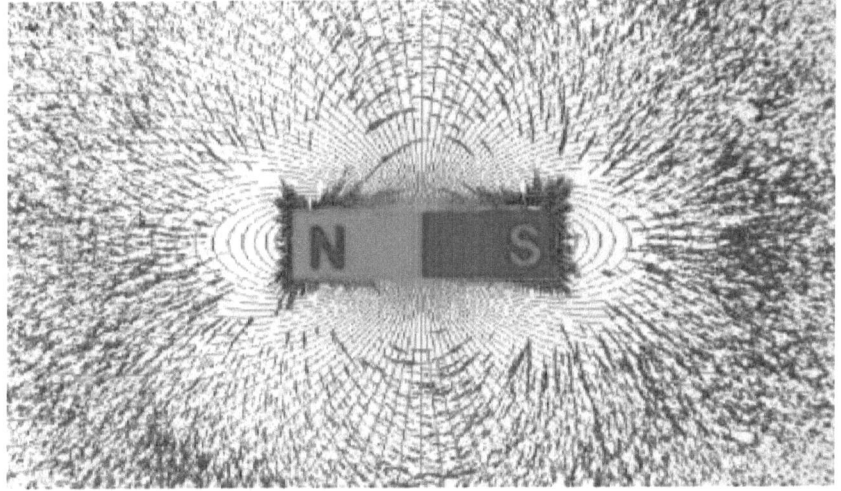

The two velocity components (which don't have to be velocity if this isn't a flow problem) are given by:

$$u = \frac{\partial \phi}{\partial x} = \frac{\partial \psi}{\partial y}$$
$$v = \frac{\partial \phi}{\partial y} = -\frac{\partial \psi}{\partial x} \tag{3.3}$$

One way of visualizing this is particles pulled along by the potential much like meteorites (and space debris) falling to the Earth along the lines of the gravitational field. Iron filings align themselves with the potential as shown in the first figure. The streamlines in the second figure are everywhere perpendicular to the potential.

The lines of constant potential are green in this next figure and the ones of constant stream function are red:

This particular solution is called a *doublet*, which is a coincident combination of a source plus a sink. Notice that the red and green lines are perpendicular and always meet at a right angle (90°). This is the graphical expression of Equation 3.3.

## Complex Combination

What does this have to do with complex variables? It's quite convenient to combine the potential and stream function into a single complex variable, which greatly simplifies many calculations:

$$w = \phi + i\psi \tag{3.4}$$

A Few Common Solutions

The simplest solution is a uniform horizontal flow to the right:

$$w = Qx + iQy = Q(x + iy) \tag{3.5}$$

In Equation 3.5 $Q$ is the strength (or magnitude) of the flow. For a uniform horizontal flow to the left, replace $Q$ with $-Q$. A uniform upward vertical flow is:

$$w = Qy - iQx = Q(y - ix) \tag{3.6}$$

For a uniform downward vertical flow, replace $Q$ with $-Q$. A uniform flow inclined at some angle, $\theta$, is:

$$w = Q(x\cos\theta + y\sin\theta) + iQ(y\cos\theta - x\sin\theta) \tag{3.7}$$

Substitute Equation 1.5 into 3.7 and the utility of complex variables for solving potential problems jumps off the page!

$$w = Q(x + iy)e^{-i\theta} \tag{3.8}$$

A source is most conveniently expressed in polar coordinates:

$$w = Q(\ln r + i\theta) \tag{3.9}$$

Here $r$ is the distance from the center and $\theta$ is the angle relative to the center. For a sink, replace $Q$ by $-Q$. A doublet is the combination of a source and sink of equal strength, brought together in the limit so that they don't cancel each other out. The doublet is very useful in constructing fields and is conveniently expressed in mixed coordinates:

$$w = \frac{Q}{r^2}(x - iy) \tag{3.10}$$

An irrotational (i.e., free) vortex of strength $Q$ in polar coordinates is:

$$w = \frac{Q}{2\pi}(\theta - i\ln r) \tag{3.11}$$

This is sometimes called a *point* vortex. As two-dimensional analysis presumes no variation in the third dimension, this might also be called a *line* vortex. The sign of $+Q$ is counter clockwise and $-Q$ is clockwise.

All of the preceding potentials and stream functions (elementary and combined) can be found in the online archive in folder examples\2D-flow. The following is an excerpt of the C source code (potflow.c):

```
typedef struct{double a,p,q,r,s,t,u,v,w,x,y,z;}vars;
vars CartesianToCylindrical(double x,double y)
   {
```

```
    static vars s;
    memset(&s,0,sizeof(s));
    s.x=x;
    s.y=y;
    s.r=hypot(x,y);
    if(s.r>DBL_EPSILON)
       s.theta=atan2(y,x);
    return(s);
}
vars UniformFlow(double angle,double x,double y)
{
    static vars s;
    memset(&s,0,sizeof(s));
    s=CartesianToCylindrical(x,y);
    s.u=cos(angle);
    s.v=sin(angle);
    s.psi=x*s.u+y*s.v;
    s.phi=y*s.u-x*s.v;
    return(s);
}
vars PointSource(double x,double y)
{
    double pi2;
    static vars s;
    memset(&s,0,sizeof(s));
    s=CartesianToCylindrical(x,y);
    if(s.r>DBL_EPSILON)
       {
       pi2=M_PI*2.;
       s.phi=log(s.r)/pi2;
       s.psi=s.theta/pi2;
       s.u=x/pow(s.r,2)/pi2;
       s.v=y/pow(s.r,2)/pi2;
       }
    return(s);
}
vars PointDoublet(double x,double y)
{
    static vars s;
    memset(&s,0,sizeof(s));
    s=CartesianToCylindrical(x,y);
    if(s.r>DBL_EPSILON)
       {
       s.phi=x/pow(s.r,2);
       s.psi=-y/pow(s.r,2);
       s.u=(pow(y,2)-pow(x,2))/pow(s.r,4);
       s.v=-2.*x*y/pow(s.r,4);
       }
    return(s);
```

```
    }
    vars CylindricalVortex(double x,double y)
    {
    double pi2;
    static vars s;
    memset(&s,0,sizeof(s));
    s=CartesianToCylindrical(x,y);
    if(s.r>DBL_EPSILON)
       {
       pi2=M_PI*2.;
       s.phi=-s.theta/pi2;
       s.psi=-log(s.r)/pi2;
       s.u=-y/pow(s.r,2)/pi2;
       s.v=x/pow(s.r,2)/pi2;
       }
    return(s);
    }
```

The structure (vars) facilitates passing the various parameters. The element q is the potential and p is the stream function. By changing the conditional compilation statements (#if... #elif... #endif) we can select between the several solutions listed previously. The preceding code is implemented using real variables. The same functions are implemented using complex variables in cmplxflo.cpp, located in the same folder. The results are the same either way.

## Superposition

Perhaps the most useful aspect of potential solutions to Laplace's equation is that we can add them up to make many combinations. This is possible because Laplace's equation is linear; whereas, the Navier-Stokes[8,9] equation is not. For example, we can add a uniform flow and point source to produce a flow past a half-body (like an infinitely long torpedo with a rounded nose). Flow past a sphere is simply a uniform flow plus a doublet. Flow past a Rankine body[10,11] is a uniform flow plus a source and a sink, separated by some distance.

## Stagnation

The simplest form of stagnated flow is against an infinite wall or plane. In two-dimensions this can be produced by a source or sink at $y=+H$ and an identical one at $y=-H$. At $y=0$ the two identical and opposite effects cancel out,

---

[8] The governing partial differential equation of viscous fluid flow named after Claude Navier and George Stokes.
[9] Claude Louis Marie Henri Navier (1785-1836) French engineer and physicist who made great advances in the study of fluid flow.
[10] A Rankine body looks like a football with oval, rather than elliptical, ends.
[11] William John Macquorn Rankine (1820-1872) Scottish engineer, physicist, and mathematician, who developed the theory of the steam engine.

producing an infinite plane where there is no flow. Set STAGNATION to 1 in potflow.c to produce the following:

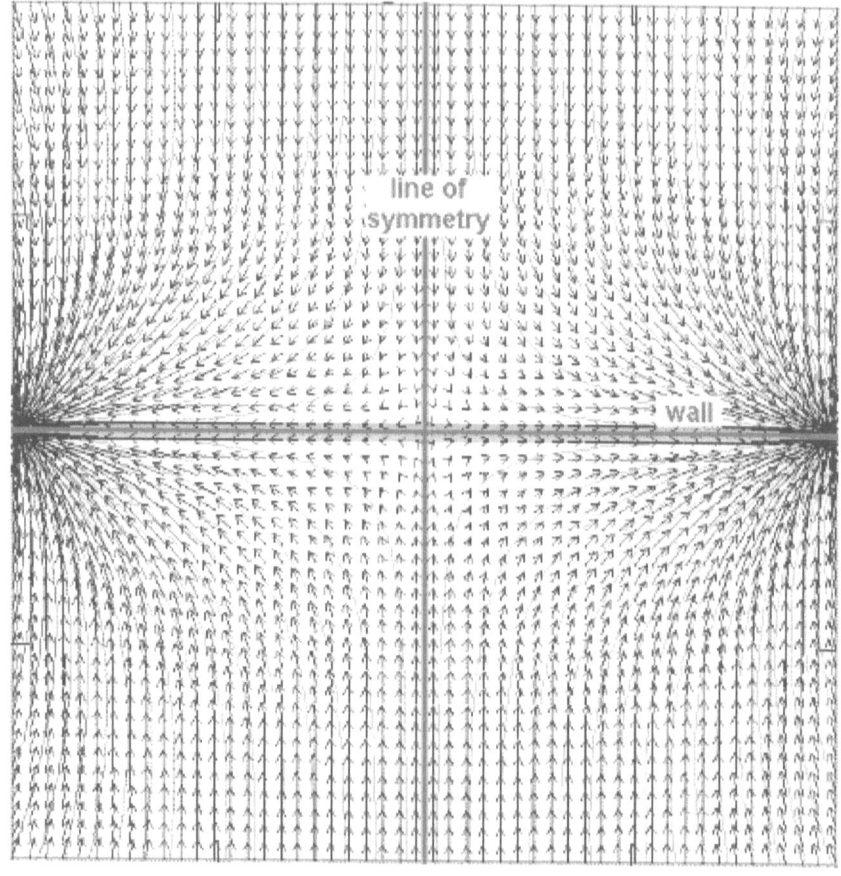

**Stagnation Flow**

## Flow Past a Cylinder

Flow past a cylinder is a uniform flow plus a doublet. If we add a vortex, this produces circulation (i.e., rotation), as shown in this next figure:

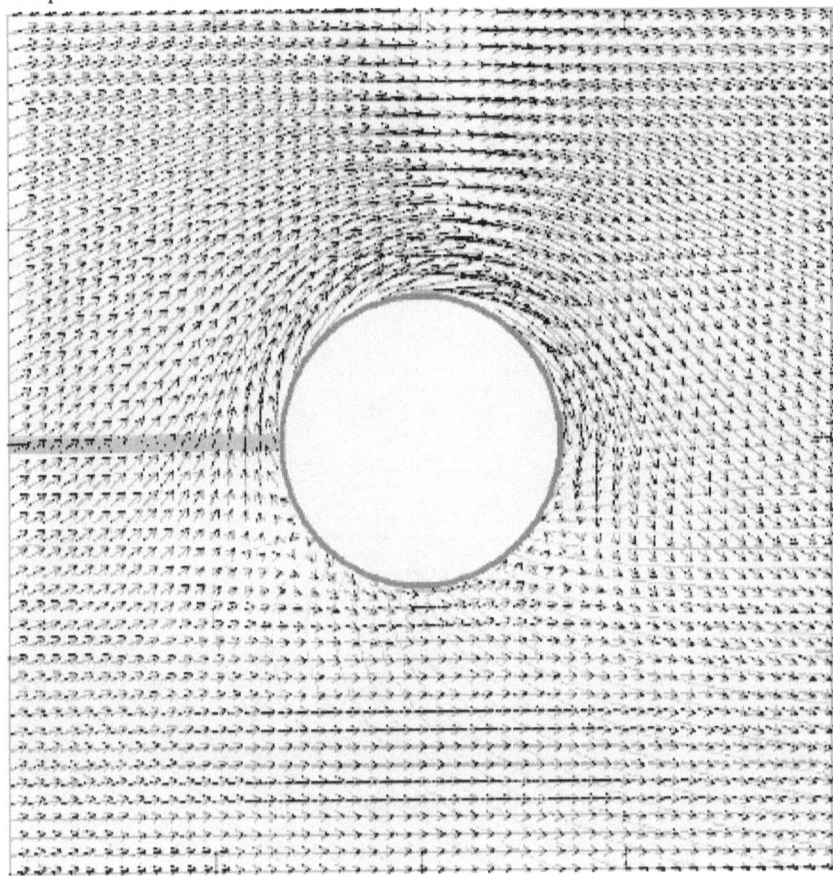

## Rankine Body

Uniform flow over a source and sink, separated by a finite difference produces what is called a Rankine body, which is sort of an elliptical shape, but more oval on the ends.

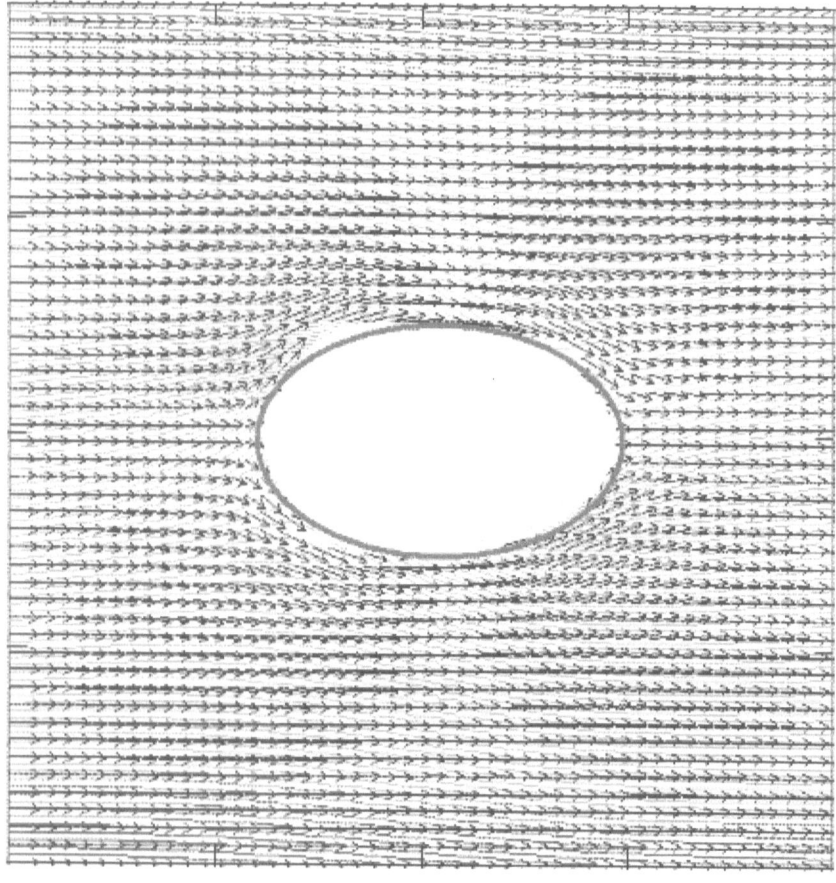

**Flow Past a Rankine Body**

These superimposed combinations are easily generated in code:
```
vars FlowPastHalfBody(double a,double x,double y)
  {
  static vars s,t;
  t=UniformFlow(0.,x,y);
  s=PointSource(x,y);
  t.phi+=a*s.phi;
  t.psi+=a*s.psi;
  t.u+=a*s.u;
  t.v+=a*s.v;
  return(t);
  }
vars FlowPastCylinder(double a,double x,double y)
  {
  static vars s,t;
```

```
    t=UniformFlow(0.,x,y);
    s=PointDoublet(x,y);
    t.phi+=a*s.phi;
    t.psi+=a*s.psi;
    t.u+=a*s.u;
    t.v+=a*s.v;
    return(t);
    }
vars FlowPastCylinderWithCirculation(double a,double
    b,double x,double y)
    {
    static vars s,t;
    t=UniformFlow(0.,x,y);
    s=PointDoublet(x,y);
    t.phi+=a*s.phi;
    t.psi+=a*s.psi;
    t.u+=a*s.u;
    t.v+=a*s.v;
    s=CylindricalVortex(x,y);
    t.phi+=b*s.phi;
    t.psi+=b*s.psi;
    t.u+=b*s.u;
    t.v+=b*s.v;
    return(t);
    }
vars FlowPastRankineBody(double a,double b,double
    x,double y)
    {
    static vars s,t;
    t=UniformFlow(0.,x,y);
    s=PointSource(x+b,y);
    t.phi+=a*s.phi;
    t.psi+=a*s.psi;
    t.u+=a*s.u;
    t.v+=a*s.v;
    s=PointSource(x-b,y);
    t.phi-=a*s.phi;
    t.psi-=a*s.psi;
    t.u-=a*s.u;
    t.v-=a*s.v;
    return(t);
    }
```

## Corner, Step, and Wedge Flow

Varying the exponent in the following equation will produce a variety of interesting flow patterns:

$$w = z^n \qquad (3.12)$$

Setting n=2 produces flow around a corner, n=3 flow over a wedge, and n=2/3 flow over a step. This next figure shows the four-corner flow:

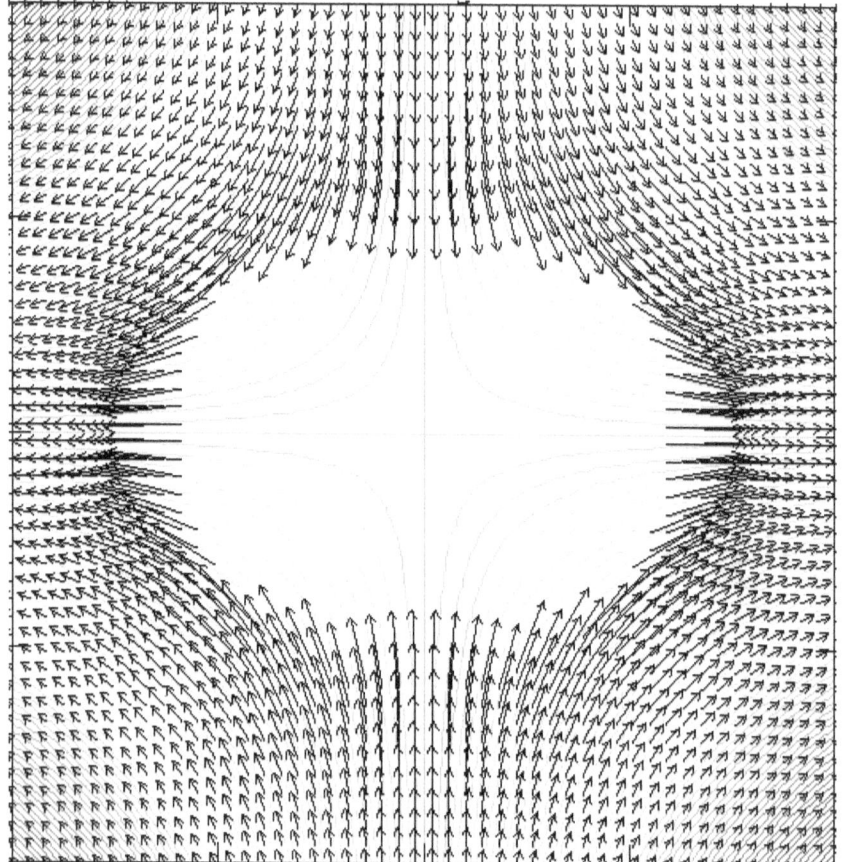

This next figure is flow over a wedge:

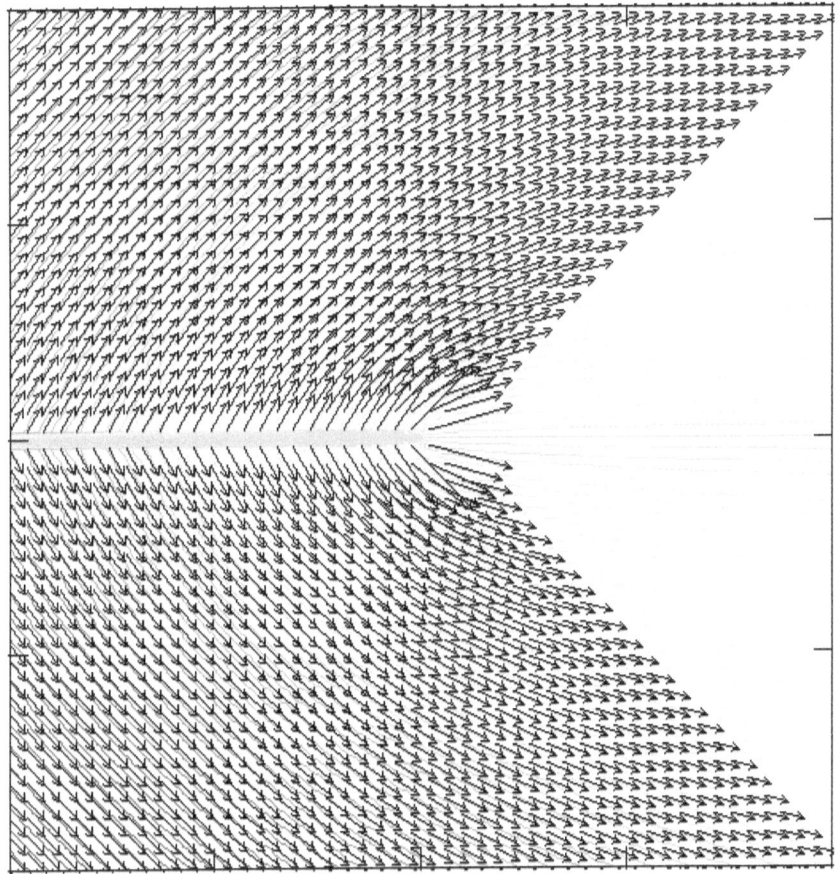

Inclined Flow Over a Plate

You will see this interesting example in many texts on potential flow and conformal mapping. Not only is this a good example of what you can do with complex variables, it's an excellent example of why I have written this and similar books; and that is, to tell the rest of the story. While this example appears in many textbooks and on many web pages, you are not likely to find the actual solution; rather, you will find a few cryptic and incomplete formulas that don't produce the pictures. That's because those authors have probably never done the calculations themselves and merely copy the pictures from some previous text, wave their hands, and shout *voila!*

This problem is a modification of inclined flow over a cylinder. The coordinates are transformed (through conformal mapping, which we will discuss later) in order to squash the cylinder into a plate. The transformation is:

$$\zeta = z + \frac{1}{z} \qquad (3.13)$$

Seems simple enough, but wait... they forgot to mention that what you really need is:

$$z = \frac{\left(\zeta \pm \sqrt{\zeta^2 - 4}\right)}{2} \qquad (3.14)$$

The hand waving isn't over yet. They also forgot to tell you that for $x<0$, you must take the absolute value of $x$ and also change the sign of $y$. Oh, yes, and there's another little thing... for $x<0$ you must change the signs on both φ and ψ. Here's the code that actually does work:

```
complex Plate(double angle,double L,complex z)
  {
  complex q;
  static vars s;
  if(z.re>=0.)
    q=z;
  else
    {
    q.re=fabs(z.re);
    q.im=-z.im;
    }
  s=UniformFlow(angle,(q+sqrt(q*q-L*L/4.))/2.);
  if(z.re<0.)
    return(-s.w);
  return(s.w);
  }
vars FlowOverPlate(double angle,double L,complex z)
  {
  double delta=0.0001;
  complex p1,p2;
  static vars s;
  p1=Plate(angle,L,z-delta/2.);
  p2=Plate(angle,L,z+delta/2.);
  s.w=(p1+p2)/2.;
  s.v.re=(p2.re-p1.re)/delta;
  s.v.im=(p1.im-p2.im)/delta;
  return(s);
  }
```

The first function just returns the potential and stream function, using the inclined uniform flow procedure from before along with the transformation in Equation 3.14. The second function calculates the two velocity components using finite differences and Equation 3.3. Only two evaluations are needed to get both velocity components, one from the potential and the other from the stream function. The proof is in the following figure:

There's a batch file to compile the source code (_compile_cmplxflo.bat) as well as a second to display the output with TP2 (_plot_cmplxflo.bat). The velocity vectors are illustrated in this next figure, which shows that you can use finite differences in this way. This is how the velocity field in the figure on page iv is generated from the complicated transformation that produces the hyperbolic shape of a natural draft cooling tower shell.

Everyone has seen these majestic devices, but most people don't realize that the shape has nothing to do with fluid mechanics. The shape is purely structural, to minimize the amount of steel and concrete. Another interesting aspect of the natural draft cooling tower shell design is that it is a surface of minimum resistance, which is what you naturally get by pulling a soap bubble between two circles. This is also how you get the ideal transition between a round and rectangular duct.

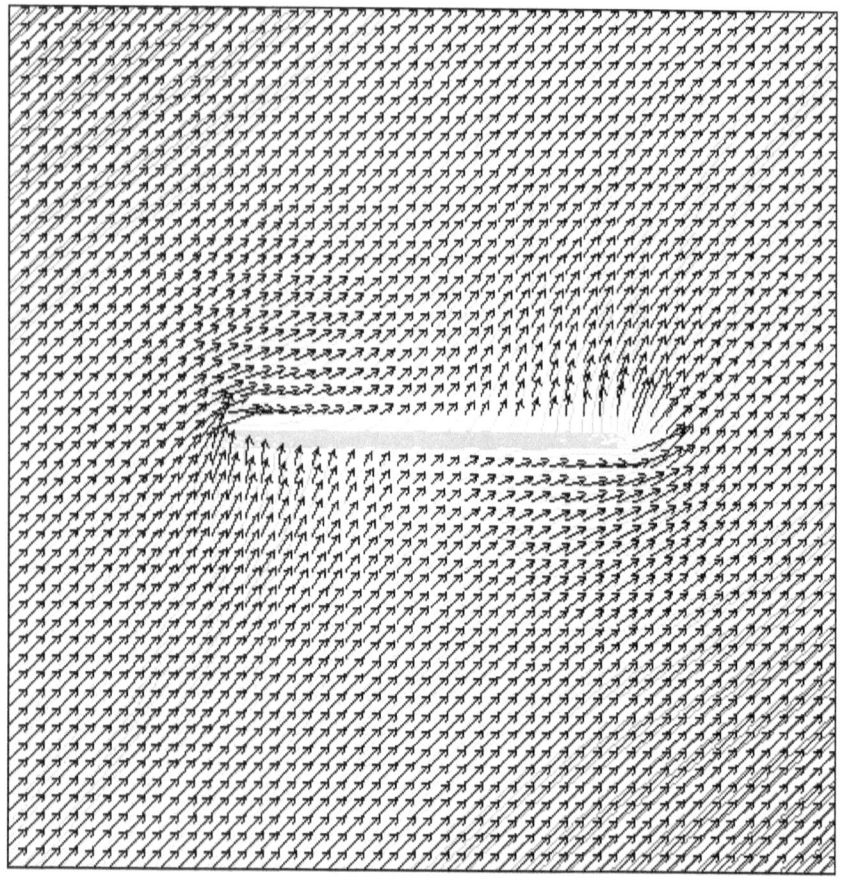

Flow out of a Nozzle

The last two-dimensional potential example we will consider looks like flow out of a nozzle. It's more general application is the electrostatic field produced by the two charged plates of a capacitor. The transformation is:

$$\zeta = e^z + z \tag{3.15}$$

Again, we want the opposite of this transform for the purposes of coding:

$$z = \zeta - W(\zeta) \tag{3.16}$$

In this formula $W(\zeta)$ is the Lambert function, which is described in Appendix B. The results are shown in the following figure:

We will explore more complicated potentials after introducing three-dimensional objects.

## Chapter 4. 3D Potentials

There are far more interesting two- than three-dimensional potential solutions. While there are infinitely many 3D solutions to Laplace's equation, only a very few of these have any meaningful stream functions or associated velocity fields that can be visualized. These few cases are mostly in spherical coordinates, though even fewer cylindrical ones exist. Laplace's equation in 3D Cartesian and spherical coordinates $(r, \theta, \xi)$, this becomes:

$$\frac{\partial^2 \phi}{\partial x^2} + \frac{\partial^2 \phi}{\partial y^2} + \frac{\partial^2 \phi}{\partial z^2} = 0$$

$$\frac{1}{r^2}\frac{\partial\left(r^2 \frac{\partial \phi}{\partial r}\right)}{\partial r} + \frac{1}{r^2 \sin\theta}\frac{\partial\left(\sin\theta \frac{\partial \phi}{\partial \theta}\right)}{\partial \theta} + \frac{1}{r^2 \sin^2\theta}\frac{\partial^2 \phi}{\partial \xi^2} = 0 \quad (4.1)$$

Once we get into three dimensions, complex variables provide no simplifications or computational advantage. The parameters and solutions aren't represented in the complex plane. Still, it would be a shame not to mention these after discussing the 2D examples. To that end, I've provided some of the same solutions in examples\3D-flow\potflow.c. The simplest is a point source:

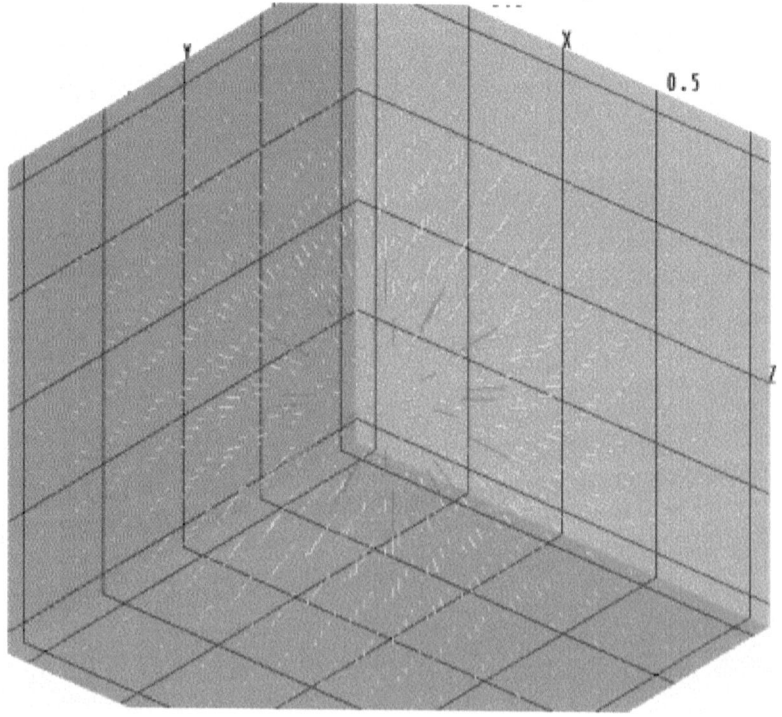

Several more are already programmed (uniform flow, point source, point doublet, line source, cylindrical vortex, spherical vortex, flow past a sphere, flow past a sphere with circulation, flow past a Rankine body, and stagnation flow). Instead of lines representing the potential and stream function, in three dimensions, these become surfaces. The following figure shows the potential iso-surfaces for stagnation flow:

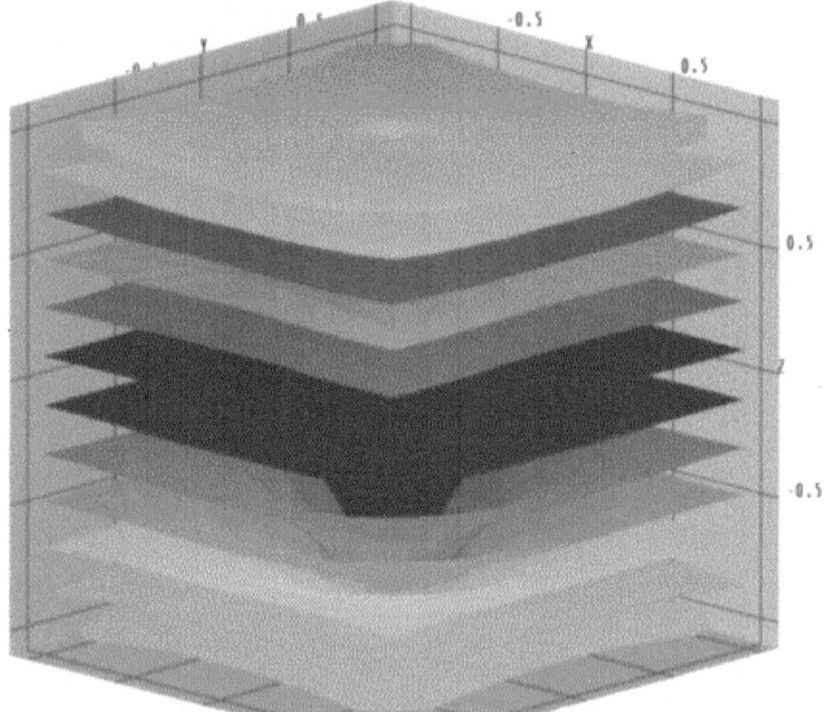

This next figure shows the iso-surfaces of the stream function for this same flow:

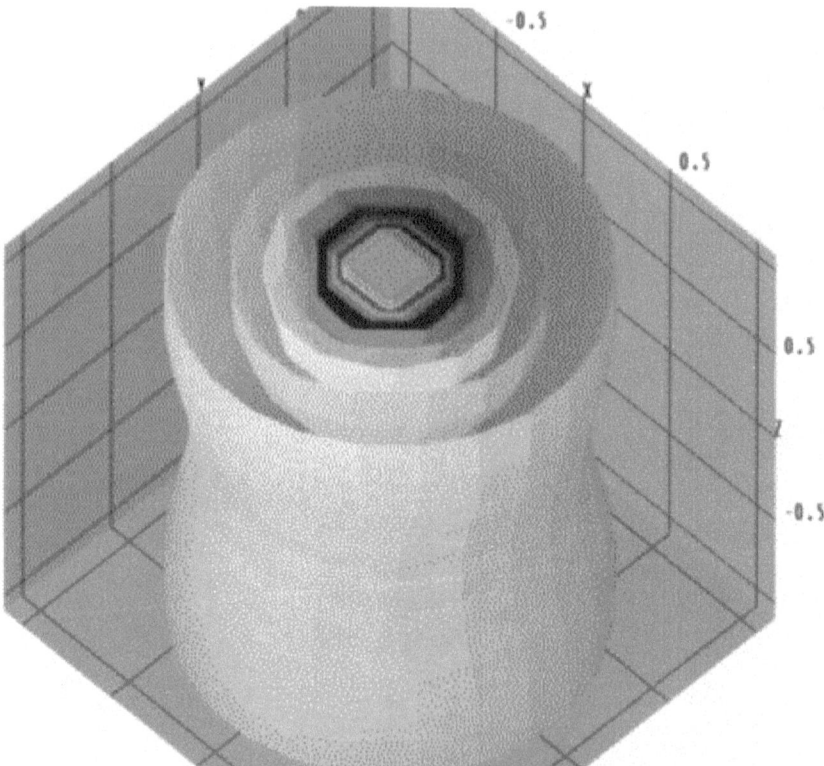

The 3D field variables are written out in a special volume format (TB3 or 3D table), which can be converted to iso-surfaces with TP2 (tools\extract shells). You can also extract contours from the 2D table files (TB2) from the previous chapter (tools\extract contours).

## Chapter 5. Conformal Mapping

Conformal Mapping is often defined as a transformation of variables that preserves angles in both magnitude and direction. It is debatable whether or not this definition is mathematically correct. As a working definition, it is utterly useless. You will find many examples of such transformations on the Web. An excellent source for information is Wolfram MathWorld®.[12] On the Conformal Mapping page you will find several examples. The first of these is $\zeta=1/z$ and is shown in the following figure:

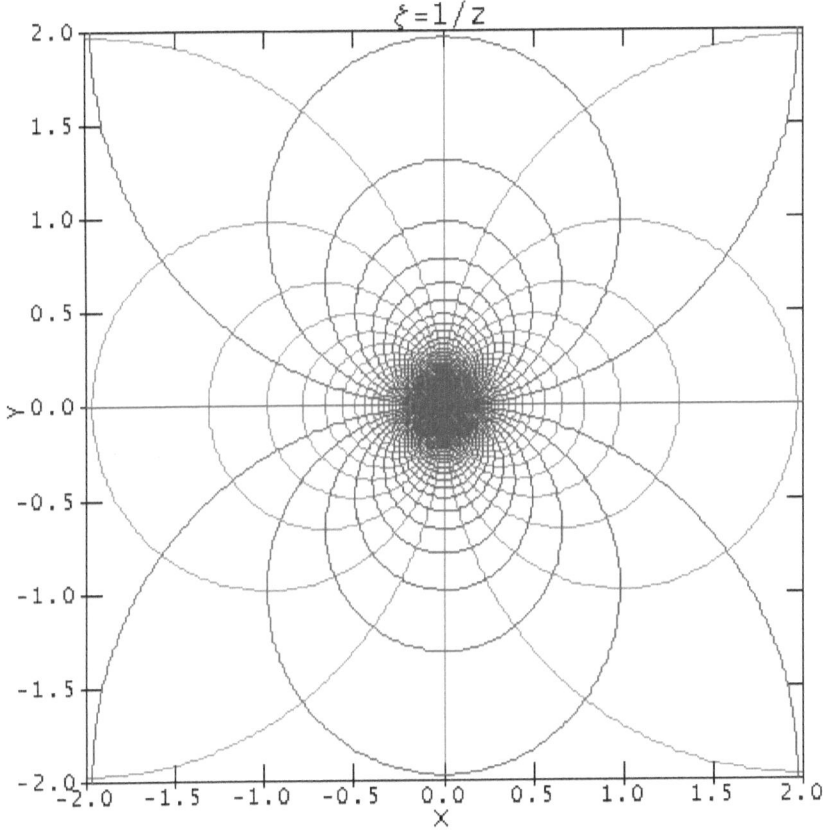

As we discussed in Chapter 3, this isn't $\zeta=1/z$; rather, it's $z=1/\zeta$.

---

[12] http://mathworld.wolfram.com/

The next example is $\zeta=sin(z)$, which is really $z=asin(\zeta)$ and shown in this next figure:

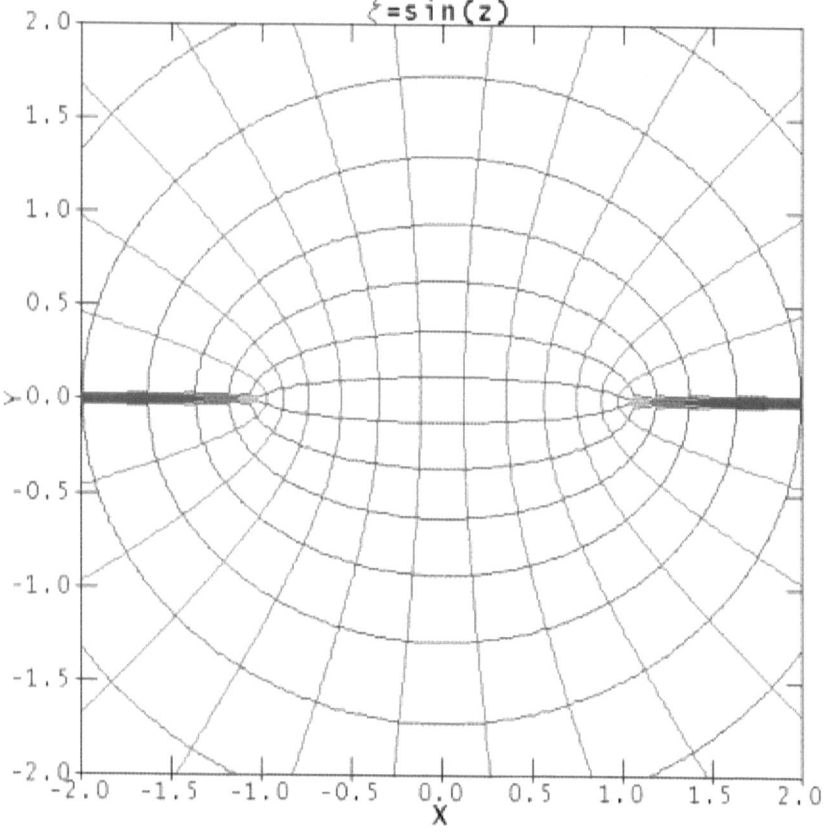

When we get to $\zeta=cos(z)$, which is really $z=acos(\zeta)$, once again, we must flip signs in order to produce the requisite figure, as in the following code and figure:

```
if(z.im>0.)
    {
    w.re=z.re;
    w.im=-z.im;
    w=acos(w);
    w.re=-w.re;
    }
else if(z.im<0.)
    w=acos(z);
else
    {
    w.re=z.re;
```

```
      w.im=0.;
    }
```

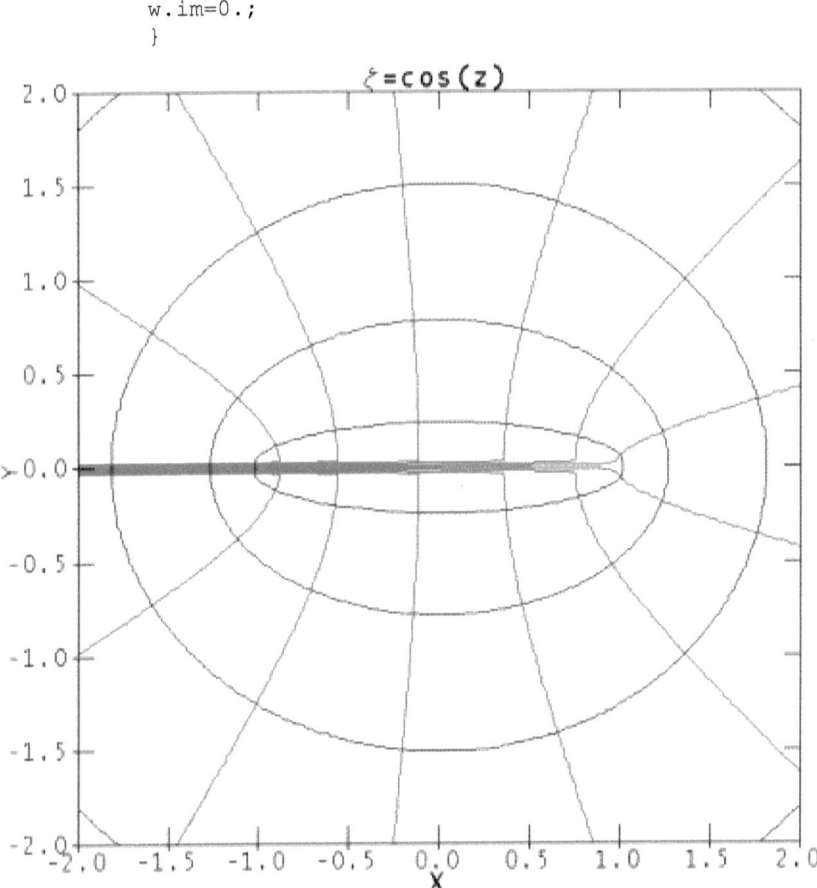

The same hand-waving exists for $\zeta=exp(z)$, which is actually $z=\ln(\zeta)$, and also requires patching up to produce this figure:

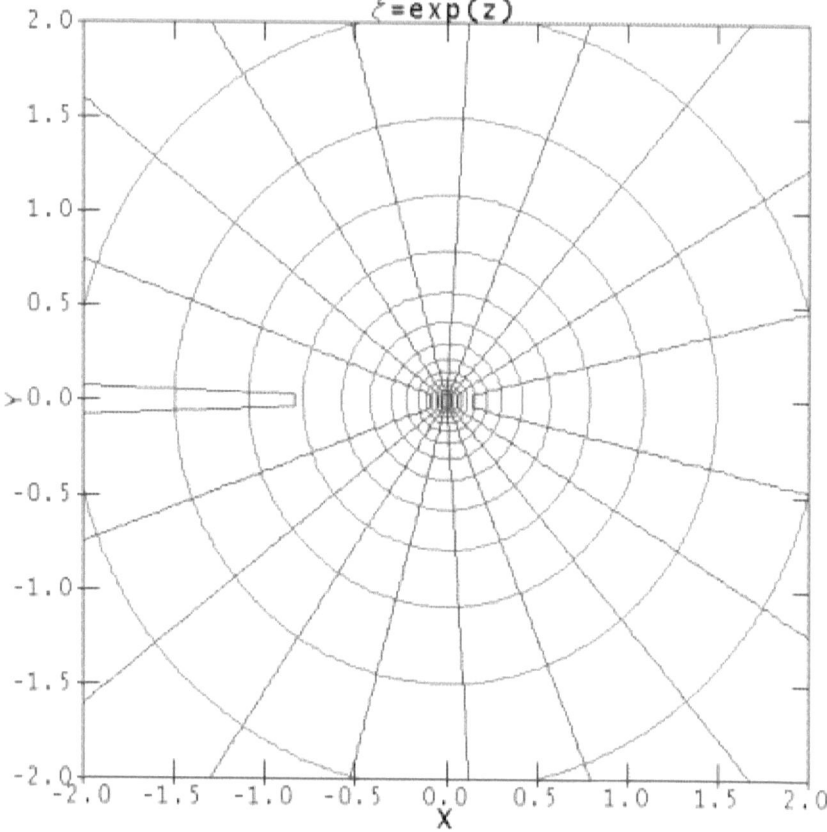

The next example is $\zeta=z^2/2$, which is really $z=\sqrt{2\zeta}$, and requires patching.

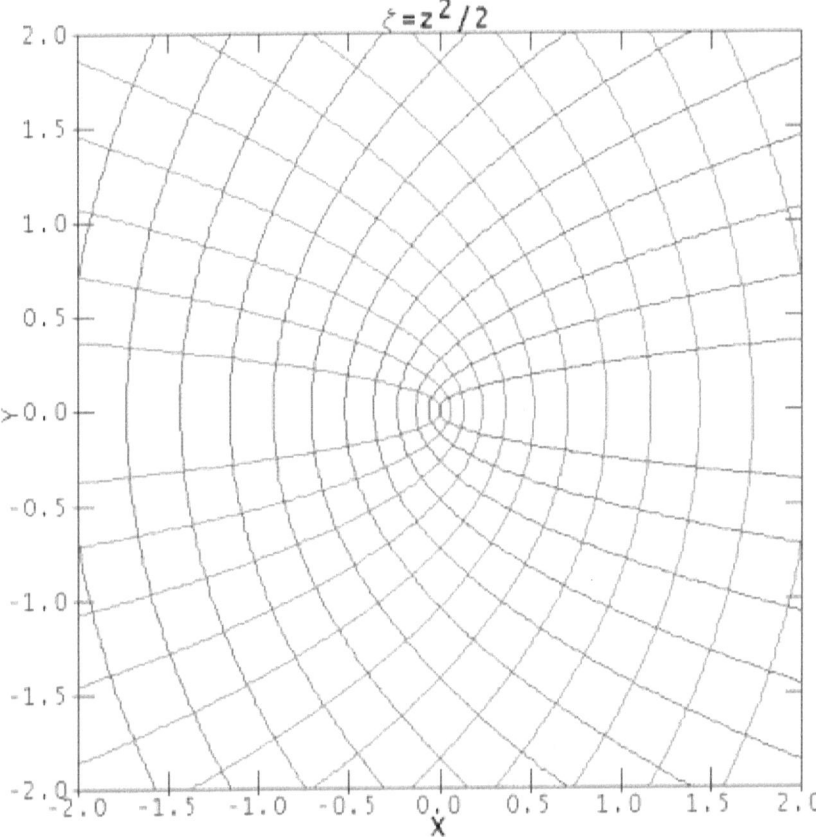

These first four examples are contours of the real and imaginary parts of the zeta ($\zeta$) function, or inverse of the transform. You can plot these same transforms along the contours using the transform variables, which you don't know beforehand. This second approach doesn't require the inverse calculations. The first representation is obtained by creating two table files (conformal1.tb2 and conformal2.tb2) containing the real and imaginary components as a 2D field plus a plot command file to combine and generate the axes (conformap.tp2). The second approach is obtained by a simple series of polygons (conformal.p2d). Both can be displayed using TP2, as described previously. All of the code to produce these and other such figures can be found in the online archive in folder examples\Conformal Mapping\conformal.cpp.

The final example in this section is the arcsin(z).

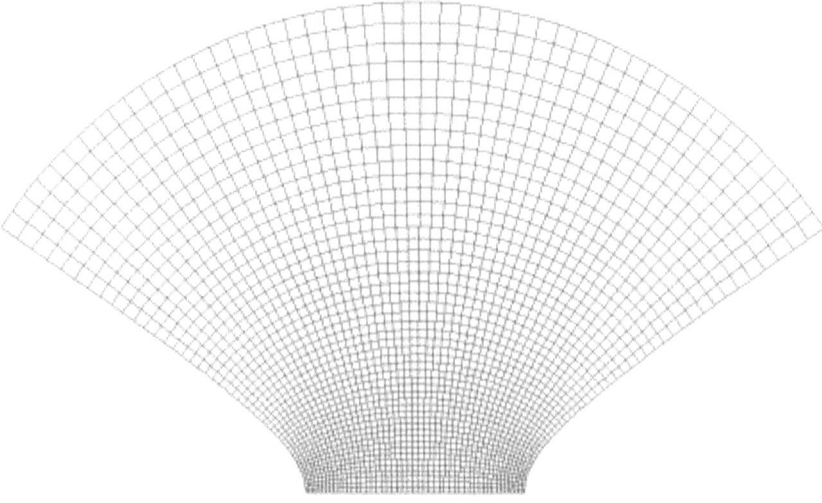

## Schwarz-Christoffel Transformation

Schwarz–Christoffel mapping is a conformal transformation of the upper half-plane onto the interior of a simple polygon.[13,14,15] This particular type of transformation can be quite useful in that many practical domains can be approximated by simple polygons. The equation defining this transformation is:

$$f(\zeta) = \int^{\zeta} \frac{Kdw}{(w-a)^{1-\frac{\alpha}{\pi}}(w-b)^{1-\frac{\beta}{\pi}}(w-c)^{1-\frac{\gamma}{\pi}}} \qquad (5.1)$$

In the equation above, $K$ is a constant and $a$, $b$, $c$ are three polygonal vertices and $\alpha$, $\beta$, $\gamma$ are the corresponding angles, $w$ represents the boundary, and $\zeta$ is the transform coordinate. While this definition is interesting, finding a new solution that isn't already available in the open literature would be worthy of a master's degree if not a doctorate. Thankfully, many simple solutions already exist. We saw five of these solutions in Chapter 4. $\zeta=\cos(z)$ maps a semi-infinite strip into something that looks like a magnet.

---

[13] Named after Hermann Schwarz and Elwin Christoffel.
[14] Karl Hermann Amandus Schwarz (1843-1921) German mathematician.
[15] Elwin Bruno Christoffel (1829-1900) German mathematician and physicist.

## Flow over a Broad-Crested Weir

We will only consider two examples of using the Schwarz–Christoffel transformation that have analytical solutions. The first of these is similar to the ideal flow over a broad-crested weir. The domain is a step down at x=0. In terms of a polygon, this is a right turn (-90° or α=-π/2) followed by a left turn (90° or β=+π/2). The third turn, closing the polygon, is a 180° (γ=π). Substituting these three angles into Equation 5.1 yields:

$$f(\zeta) = \frac{1}{\pi}\left[\sqrt{z^2-1} + \ln\left(z + \sqrt{z^2-1}\right)\right] \tag{5.2}$$

As in several previous examples, implementation is not straightforward, requiring two patches, as shown below:

```
w=(sqrt(z*z-1.)+log(z+sqrt(z*z-1.)))/M_PI;
if(z.re<0.)
    w=-w;
w-=I;
if(w.re<0.)
    if(w.im<0.)
        w.im=0.;
```

The source code can be found in folder examples\Schwarz-Christoffel file schwarz.cpp. The results are shown in the following figure:

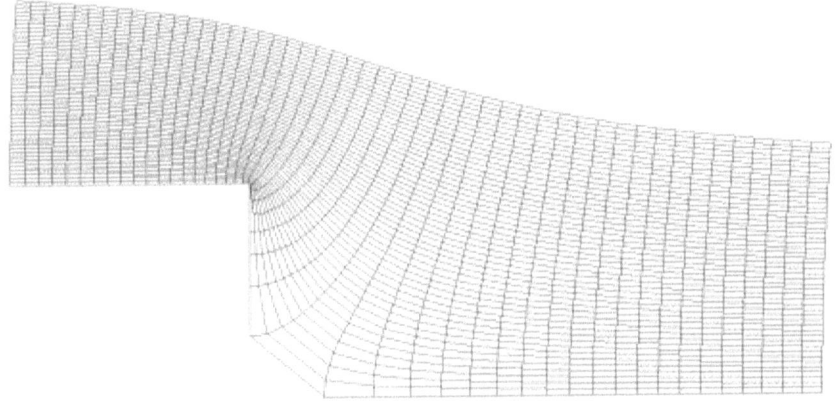

## Sharp-Crested Weir

The second example is similar to the ideal flow over a sharp-crested weir. Instead of taking an abrupt right turn at zero, we left (upward) turn at -1 (+45° or α=π/4), continue onward to zero, take a sharp downward turn (-135° or β=-3π/4), continue onward to -1, take a left (+90° or γ=+π/2), and close the polygon with a 180° (δ=π). Substituting into Equation 5.1 and integrating yields:

35

$$f(\zeta) = \frac{i(3-z)\sqrt{z}}{2} \tag{5.3}$$

This particular transformation can be implemented without patching. The code can also be found in schwarz.cpp (case=2). The results are shown below:

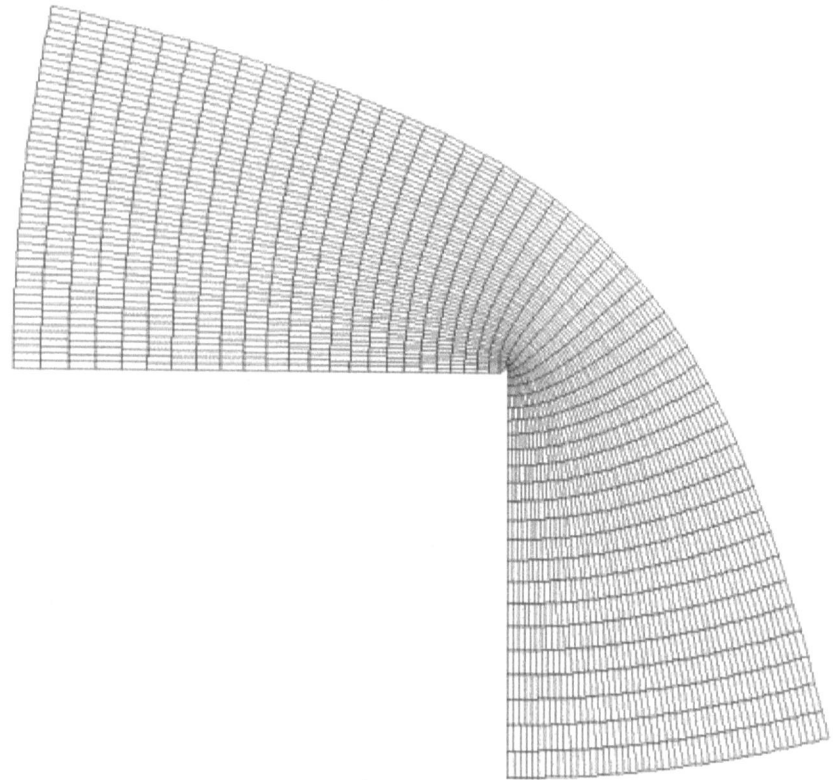

## Numerical Methods

A variety of numerical methods exist for conformal mapping. Early ones were published by Lloyd N. Trefethen (SCPACK) and more recent ones by Donald E. Marshall (Zipper). You will find many publications on the Web written by these two authors or referencing their work. Several implementations are available for download, including various source codes. Some of Trefethen's code is included in the online archive in folder examples\scpack, along with five simple examples. Case 1 is shown below:

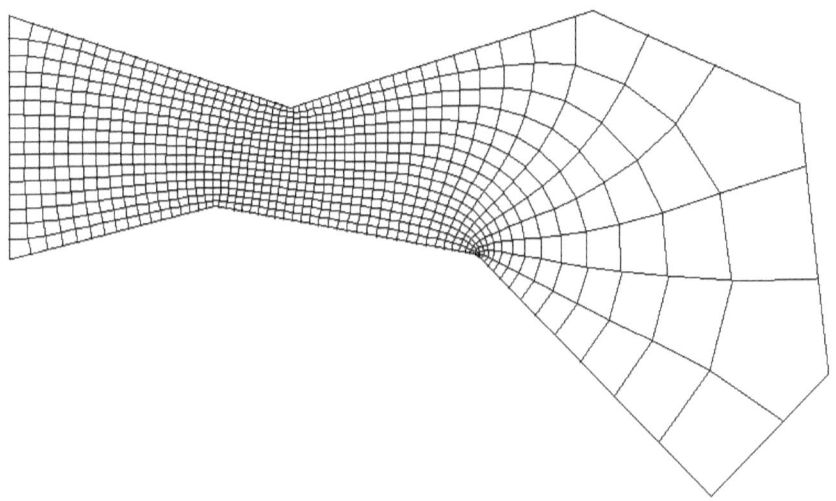

This next figure is Case 2:

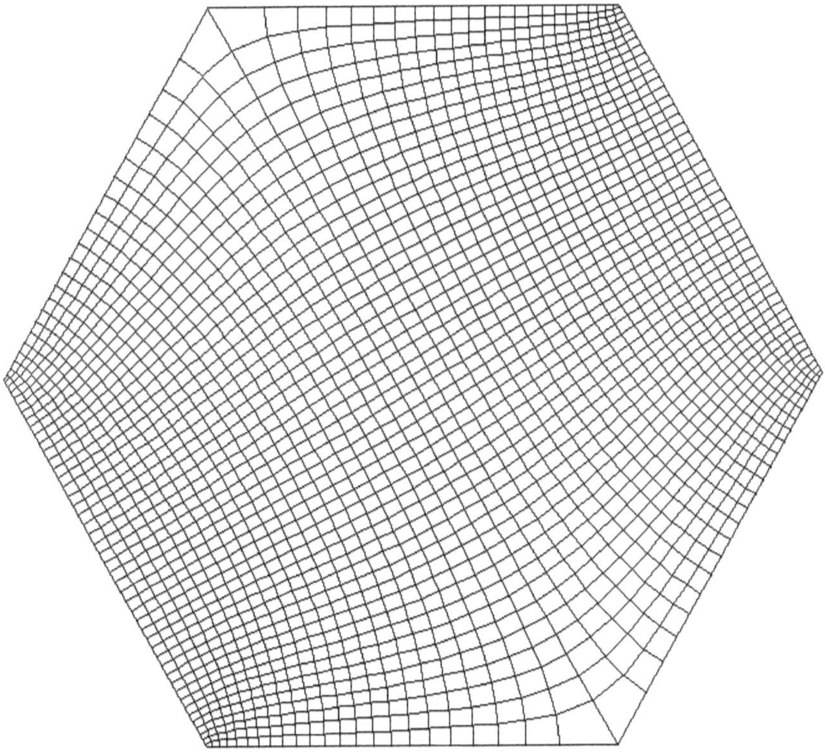

Case 3 is shown in the following figure:

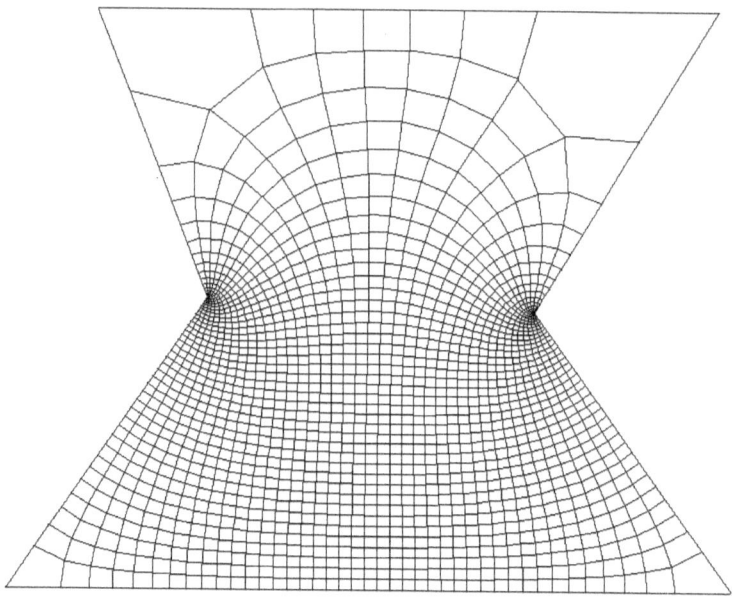

This next figure is Case 4:

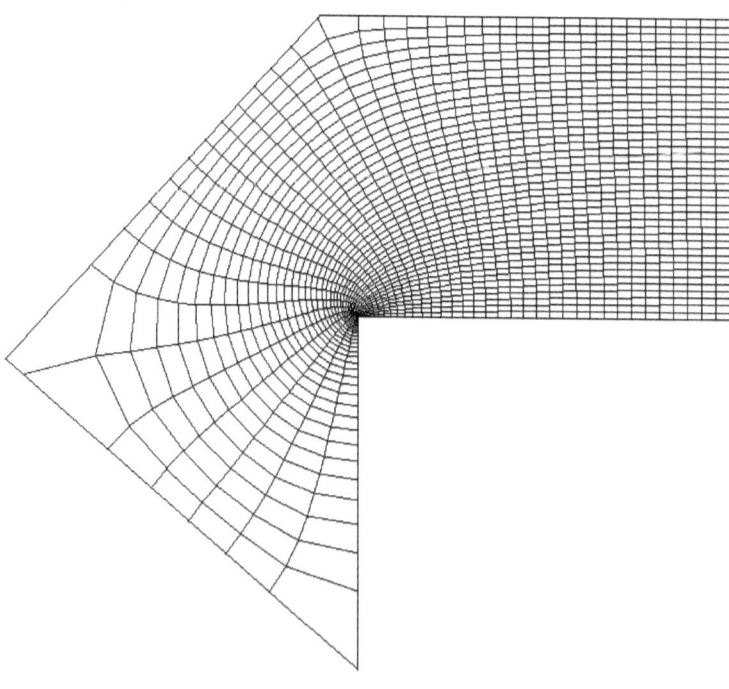

And the final example is Case 5:

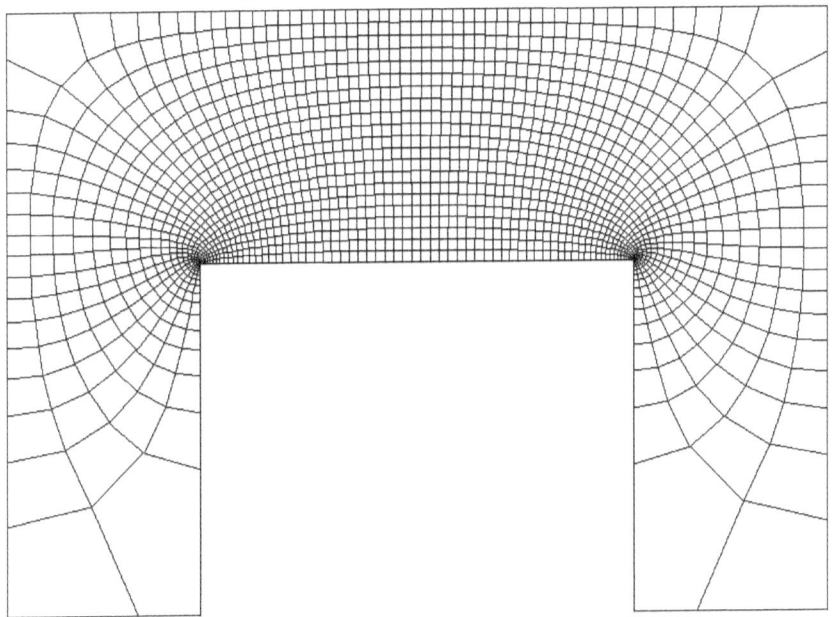

## Zipper Algorithm

The algorithm used in SCPACK code will only handle the simplest of cases. The iteration often errors out, aborting the program, or fails to converge. The method is not robust, which is discussed in several excellent articles on the Web. The zipper algorithm, as implemented by Marshall and Rohde, is much more robust.[16] The software, examples, and documents can be downloaded from Marshall's home page: https://sites.math.washington.edu/~marshall/zipper.htm

---

[16] Marshall, D. E. and S. Rohde, "Convergence of a Variant of the Zipper Algorithm for Conformal Mapping," SIAM Journal of Numerical Analysis, Vol. 45, No. 6, pp 2577–2609, 2007.

Not only is Marshall's zipper algorithm more robust, it can be applied to exterior as well as interior grids, as shown below:

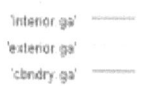

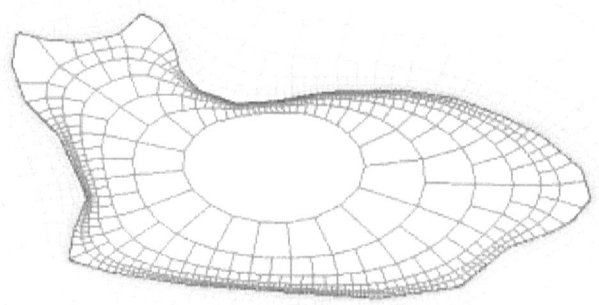

The zipper algorithm approximates the domain boundary with a series of arcs, and then links Green's Lemma through simple numerical integration to the transform like Schwarz–Christoffel (Equation 5.1). After refining the points along the boundary and solving the equations using complex variables (most conveniently done in FORTRAN), the inverse transform can be used to warp a regular pattern, as shown above for the Carlson grid.[17,18,19,20]

Marshall's software is extensive and versatile, but a little hard to follow. The terms are also unique to this narrow discipline (numerical conformal mapping). For instance, "vertices" means "domain boundary polygon points" and "prevertices" means "refined domain boundary polygon points." Multiple programs must be used to accomplish a single task. The role of each step in this process is obscure (e.g., adding extra points along the boundary and then later removing them). The half-dozen separate programs read some files (data, options, and file names) and create some output that must be fed into yet another program and so on to arrive at a final result.

---

[17] Carleson, L., "An Interpolation Problem for Bounded Analytic Functions," American Journal of Mathematics, Vol. 80, pp. 921-930, 1958.

[18] Carleson, L., "Interpolations by Bounded Analytic Functions and the Corona Problem," Annals of Mathematics, Vol. 76, pp. 547-559, 1962.

[19] Carleson, L., "The Extension Problem for Quasi-Conformal Mappings," in *Analysis*, Academic Press, New York, 1974.

[20] Carleson, L. and T. W. Gamelin, Complex Dynamics, Springer-Verlag, New York, 1993.

For convenience, I have combined all of these various programs into one. This combined program (ZIPPER.FOR) reads a single file containing the domain boundary polygon points (oriented counter-clockwise) plus a single interior point and produces a single output file containing the results (ZIPPER.P2D). It deletes all of the intermediate files upon completion. You can launch the program with the name of the boundary polygon file or simply drop it on the executable. I have provided five examples Marshall's "blob", Africa, Australia, Lake Michigan, and Lake Superior. You will find everything in the online archive in folder examples\zipper. Examples are easy to create and manage with free tools like TP2 and the polygon editor available on my website.

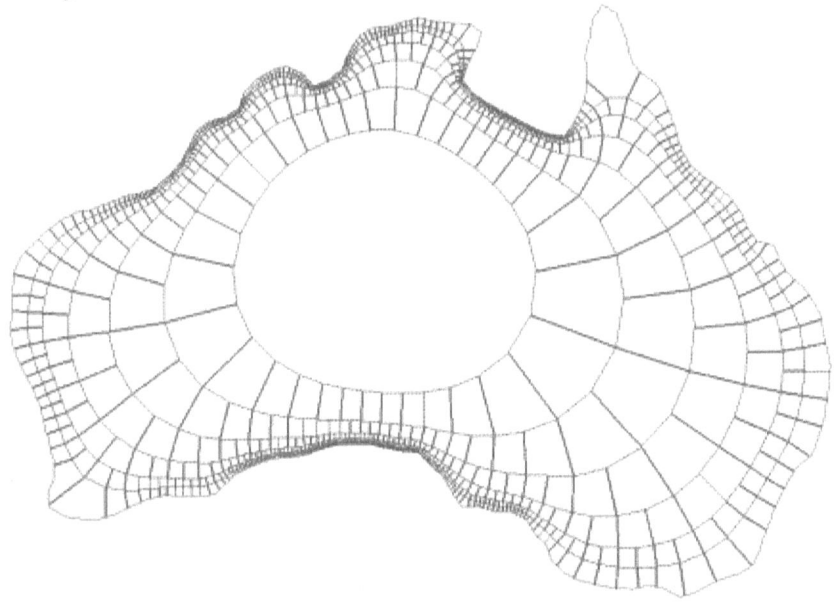

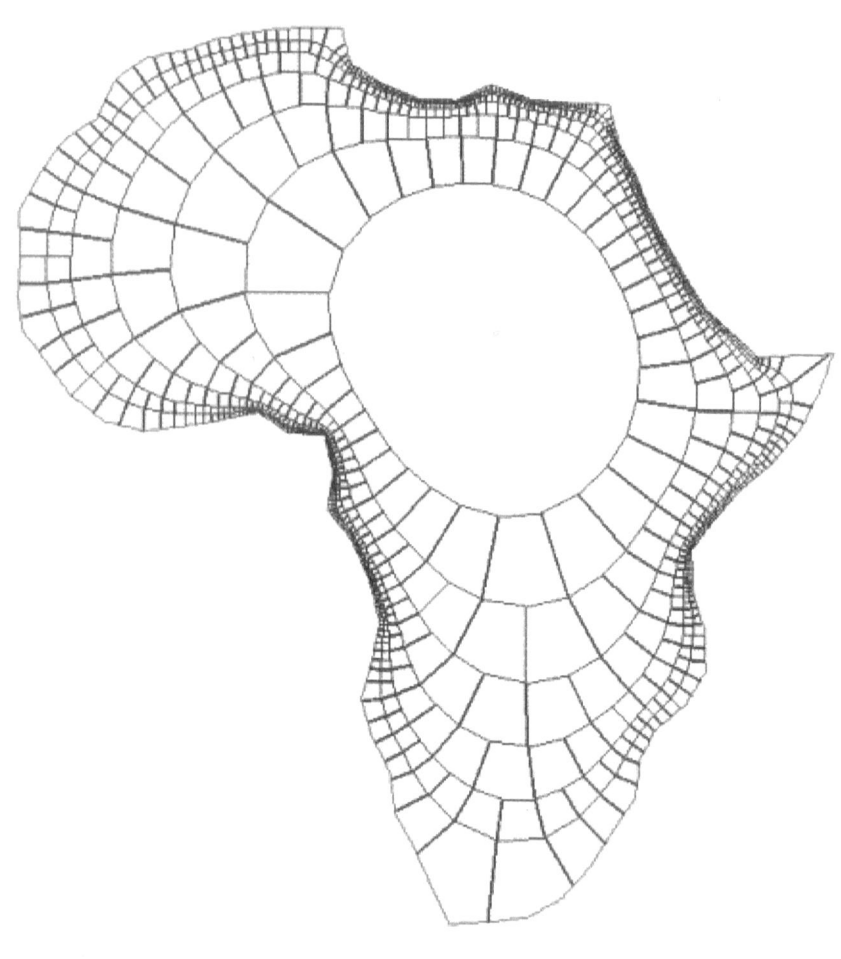

## Chapter 6. Polynomial Roots

Roots of a polynomial are often the first exposure to complex variables. Indeed, if such a polynomial were conceptualized as a surface, the roots would be the locations where the surface intersects the zero plane. Consider perhaps the simplest example of a polynomial ($y=x^2-2*x+2$) having the roots $1\pm i$. The residual or norm(y) equal to the sum of the real and imaginary components produces the following surface:

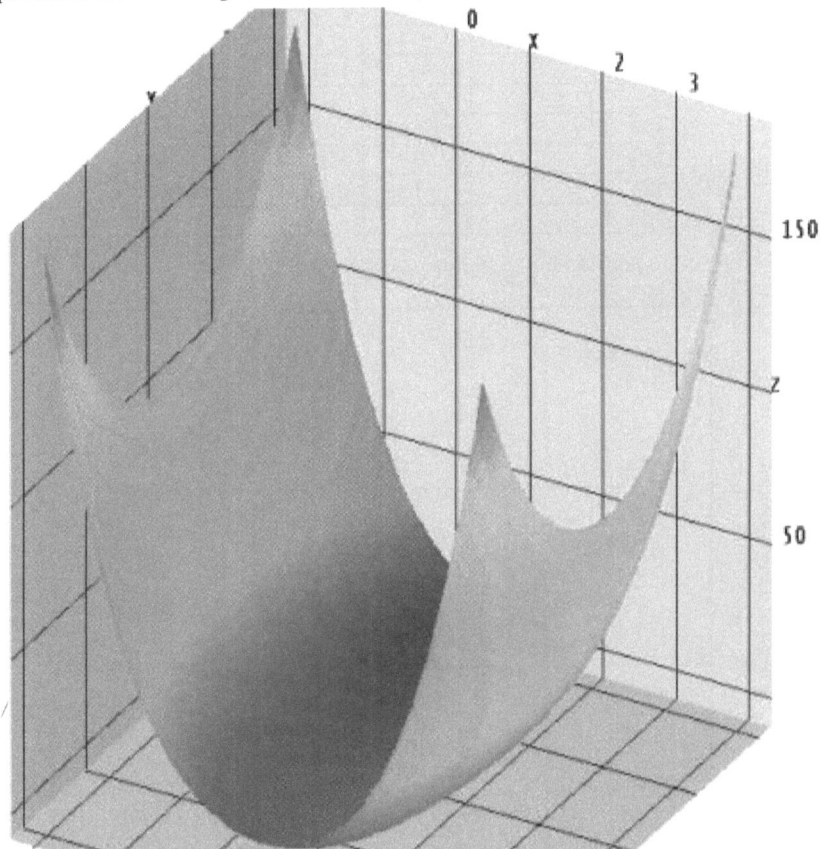

What would happen if we were to try and solve this problem using the secant method, ignoring the fact that the complex plane has two dimensions? It's a two-dimensional problem when cast in this form so we might not expect a one-dimensional solution procedure to work at all, or at least not efficiently. The code can be found in folder examples\roots in file root1.cpp.

Starting at (0.5,0.5) the results are:

| iter | x | | y | |
|---|---|---|---|---|
| step | real | imag | real | imag |
| 1 | 0.5 | 0.5 | 1 | -0.5 |
| 2 | -0.5 | 1 | 2.25 | -3 |
| 3 | 0.94 | 0.58 | 0.6672 | -0.0696 |
| 4 | 1.17343 | 0.771805 | 0.434394 | 0.267705 |
| 5 | 0.950002 | 1.0744 | -0.15184 | -0.10744 |
| 6 | 1.01341 | 0.996397 | 0.007373 | 0.026723 |
| 7 | 1.00057 | 1.00018 | -0.00037 | 0.001144 |
| 8 | 1 | 0.999996 | 8.35E-06 | 4.68E-07 |
| 9 | 1 | 1 | 9.03E-10 | -2.34E-09 |
| 10 | 1 | 1 | -2.22E-15 | 5.33E-15 |
| 11 | 1 | 1 | 0 | 0 |

The solution path is shown in this next figure:

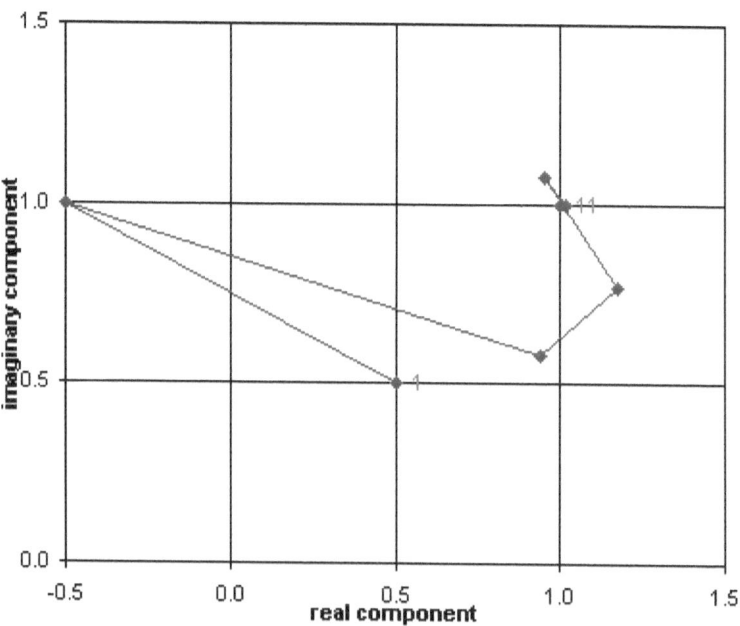

The real and imaginary components of the residual are shown in the next figure and the number of iterations required to reduce the residual to some level depend on the starting value is shown in the one after that.

44

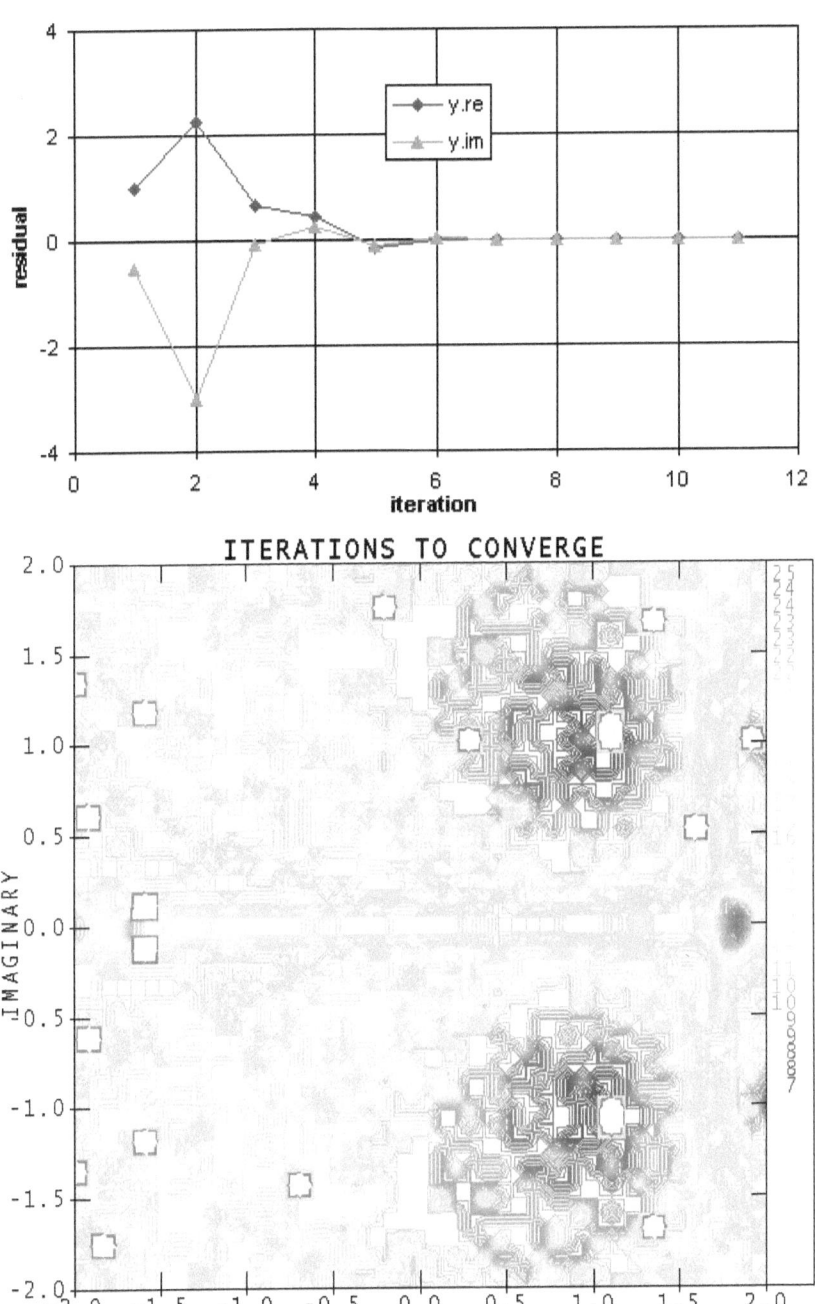

The real and imaginary components of $w=(z-r_1)*(z-r_2)*(z-r_3)*(z-r_4)$ are shown in the following figure along with the zero plane. These intersect at the four roots.

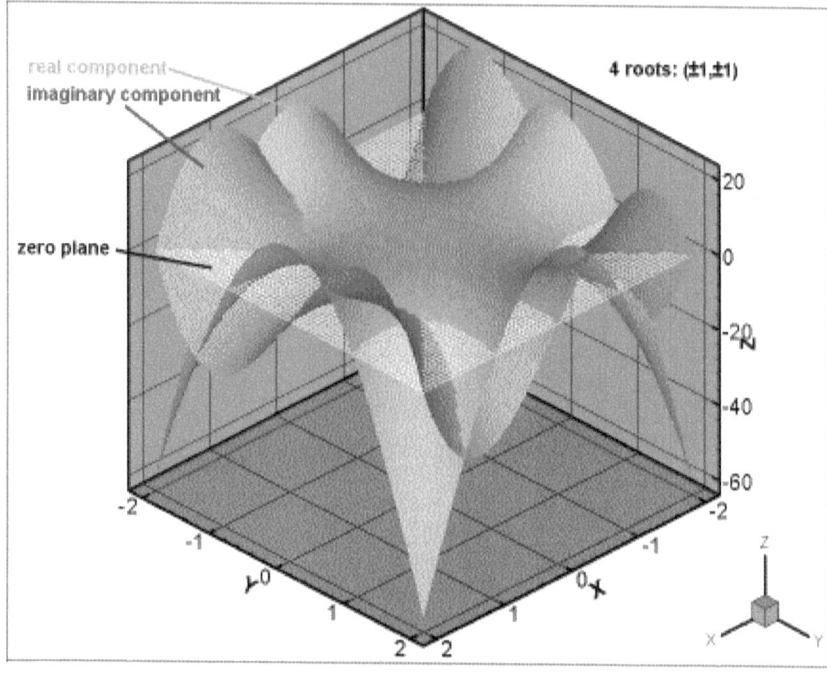

The magnitude (square root of the sum of the components squared) of the residual is shown in this next figure. The 4 roots are located at the bottom points:

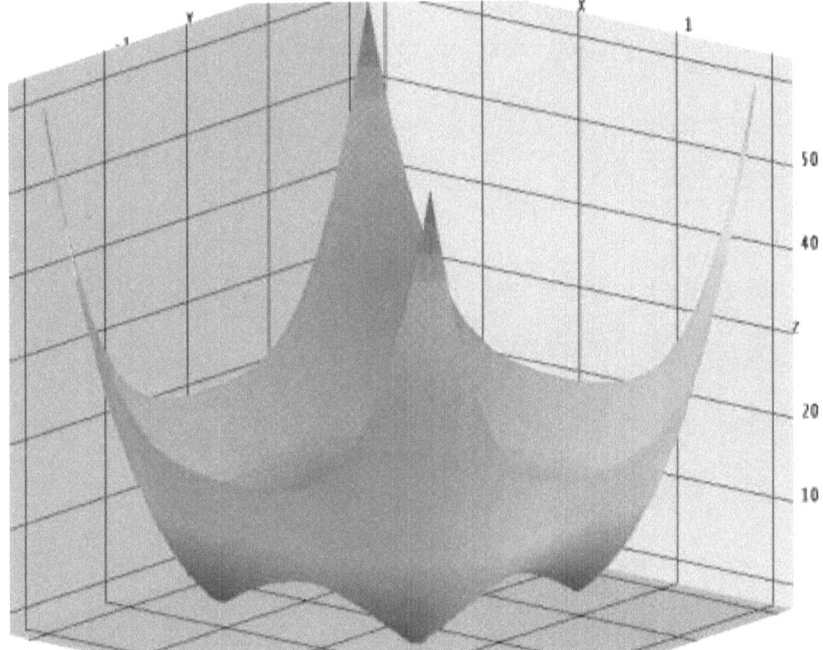

Attempting to use the secant (or Newton's) method to find these roots would be ineffective, as there is a large flat region in the center. This might work if you happened to select a starting point on one of the strongly downward sloping corners, but such doesn't form the basis for any practical method for general use.

## Bairstow's Method

The best way to find the roots (real and imaginary) of a polynomial with real coefficients is Bairstow's method.[21] The algorithm is based on the fact that any polynomial with real coefficients can be represented as a product of two smaller ones, as in the following:

$$a_0 + a_1 x + a_2 x^2 + a_3 x^3 + a_4 x^4 + a_5 x^5 + a_6 x^6 + ... = \\ (b_0 + b_1 x + b_2 x^2 + b_3 x^3 + b_4 x^4 + ...)(c_0 + c_1 x + c_2 x^2) \quad (6.1)$$

Newton's method along with polynomial division is used to find the coefficients $b_i$ and $c_i$. As is often the case with Newton's method, it may not

---

[21] This algorithm first appeared in the appendix of the 1920 book *Applied Aerodynamics* by Leonard Bairstow.

converge, depending on the starting values. This is easily remedied by trying random initial estimates within an iteration loop. The quadratic form is easily factored. The polynomial is reduced until second or third order is reached. The cubic is also easily factored. The core algorithm is as follows:

```
int Bairstow(double*a,int n,double*p,double*q)
{/* Bairstow's Method for roots */
int error=1,i,iter;
double*b,*c,cpnm1,d,deltap,deltaq,dp,dq;
b=calloc(n,sizeof(double));
c=calloc(n,sizeof(double));
for(iter=0;iter<100;iter++)
  {
  b[0]=a[0]-*p;
  c[0]=b[0]-*p;
  b[1]=a[1]-*p*b[0]-*q;
  c[1]=b[1]-*p*c[0]-*q;
  for(i=2;i<n;i++)
    {
    b[i]=a[i]-*p*b[i-1]-*q*b[i-2];
    c[i]=b[i]-*p*c[i-1]-*q*c[i-2];
    }
  cpnm1=c[n-2]-b[n-2];
  d=c[n-3]*c[n-3]-cpnm1*c[n-4];
  if(fabs(d)<FLT_MIN)
    {
    error=0;
    break;
    }
  dp=b[n-2]*c[n-3]-b[n-1]*c[n-4];
  dq=-b[n-2]*cpnm1+b[n-1]*c[n-3];
  deltap=dp/d;
  deltaq=dq/d;
  *p+=deltap;
  *q+=deltaq;
  if(fabs(deltap)+fabs(deltaq)<FLT_EPSILON)
    {
    error=0;
    break;
    }
  }
if(error==0)
  for(i=0;i<(n-2);i++)
    a[i]=b[i];
free(b);
free(c);
return(error);
}
```

A small program (bairstow.c) to implement this with random inputs can be found in the folder \examples\Bairstow. Typical output is as follows:

```
a=1,2
r=(-0.5+0i)
a=1,2,3
r=(-0.333333+0.471405i),(-0.333333-0.471405i)
a=1,2,3,4
r=(-0.60583+0i),(-0.0720852+0.638327i),(-0.0720852-
   0.638327i)
a=1,2,3,4,5
r=(0.137832+0.678154i),(0.137832-0.678154i),(-
   0.537832+0.358285i),(-0.537832-0.358285i)
a=1,2,3,4,5,6
r=(-0.670332+0i),(0.294195+0.668367i),(0.294195-
   0.668367i),(-0.375695+0.570175i),(-0.375695-
   0.570175i)
a=6.827,-0.039,-9.509,-7.005,1.942,-5.173,-
   4.564,2.39,4.604
r=(0.468748+0.989118i),(0.468748-0.989118i),(-
   0.813158+0.711306i),(-0.813158-0.711306i),(-
   0.931776+0.539861i),(-0.931776-
   0.539861i),(1.36177+0i),(0.671487+0i)
a=-6.19,1.321,0.332,7.673
r=(0.856175+0i),(-0.449722+0.860229i),(-0.449722-
   0.860229i)
a=5.141,-2.289,-1.748,-3.132
r=(0.850745+0i),(-0.704427+1.19716i),(-0.704427-
   1.19716i)
```

## Traub-Jenkins Method

Real and complex roots of a polynomial having complex coefficients are found with the Traub-Jenkins Method.[22] This algorithm is similar to Bairstow's in that a Newton iteration is used to solve for the roots. Besides the complex coefficients, the biggest difference between the two is that the Traub-Jenkins method yields a single complex root with each step. Multiple initialization attempts are also used to handle the convergence problem.

The residual (i.e., complex error associated with each provisional root) is calculated and iteratively minimized. The polynomial is also scaled in order to better estimate the root. This approach typically finds the roots in order of increasing magnitude, but this is not always the case, so don't count on it. As is often the case, convergence is roughly quadratic, making this a fast algorithm.

---

[22] Jenkins, M. A. and Traub, J. F., "A Three-Stage Variables-Shift Iteration for Polynomial Zeros and Its Relation to Generalized Rayleigh Iteration," Numerical Mathematics, Vol. 14, pp. 252–263, 1970.

The algorithm passes various modified forms of the ever-shrinking polynomial to several functions through global variables. In the earliest implementations, these arrays were static; however, in the code provided here (traubj.c), the arrays are dynamically allocated, eliminating any constraing on the degree. The core algorithm is listed below:

```
int TraubJenkins(double*a,int ndeg,double*z)
{
BOOL conv;
int i,icnt1,icnt2,ii,inx,inxi,j,n2,nn2;
double bnd,cosr,sinr,xx,xxx,yy,zi,zr;
if(ndeg<1)
   return(__LINE__);
tj.are=DBL_EPSILON;
tj.rmre=2.*M_SQRT2*DBL_EPSILON;
xx=M_SQRT1_2;
yy=-xx;
cosr=-0.06975647;
sinr=0.9975641;
tj.nn=ndeg+1;
if(fabs(a[0])>FLT_MIN||fabs(a[1])>FLT_MIN)
{
  while(1)
    {
    nn2=tj.nn+tj.nn;
    if(fabs(a[nn2-2])>FLT_MIN||fabs(a[nn2-1])>FLT_MIN)
      break;
    inx=ndeg-tj.nn+2;
    inxi=inx+ndeg;
    z[inxi-1]=z[inx-1]=0.;
    tj.nn--;
    if(tj.nn==1)
      return(0);
    }
  for(i=1;i<=tj.nn;i++)
    {
    ii=i+i;
    tj.pr[i-1]=a[ii-2];
    tj.pi[i-1]=a[ii-1];
    tj.shr[i-1]=modulus(tj.pr[i-1],tj.pi[i-1]);
    }
  bnd=ScaleFactor(tj.nn,tj.shr);
  if(bnd!=1.)
    {
    for(i=0;i<tj.nn;i++)
      {
      tj.pr[i]*=bnd;
      tj.pi[i]*=bnd;
      }
```

```
      }
   while(tj.nn>2)
      {
      for(i=0;i<tj.nn;i++)
         tj.shr[i]=modulus(tj.pr[i],tj.pi[i]);
      bnd=LowerBound(tj.nn,tj.shr,tj.shi);
      for(icnt1=1;icnt1<=2;icnt1++)
         {
         dPdH(5);
         for(icnt2=1;icnt2<=9;icnt2++)
            {
            xxx=cosr*xx-sinr*yy;
            yy=sinr*xx+cosr*yy;
            xx=xxx;
            tj.sr=bnd*xx;
            tj.si=bnd*yy;
            PolyZero2(10*icnt2,&zr,&zi,&conv);
            if(conv)
               goto next;
            }
         }
      return(__LINE__);
next:
      inx=ndeg+2-tj.nn;
      inxi=inx+ndeg;
      z[inx-1]=zr;
      z[inxi-1]=zi;
      tj.nn--;
      for(i=0;i<tj.nn;i++)
         {
         tj.pr[i]=tj.qpr[i];
         tj.pi[i]=tj.qpi[i];
         }
      }
   ComplexDivide(-tj.pr[1],-
tj.pi[1],tj.pr[0],tj.pi[0],&z[ndeg-1],&z[ndeg+ndeg-
1]);
   for(i=1;i<=ndeg;i++)
      tj.pi[i-1]=z[ndeg+i-1];
   n2=ndeg+ndeg;
   j=ndeg;
   for(i=1;i<=ndeg;i++)
      {
      z[n2-2]=z[j-1];
      z[n2-1]=tj.pi[j-1];
      n2-=2;
      j--;
      }
   return(0);
```

```
    }
    return(__LINE__);
}
```

This code can be found in the folder \examples\Traub-Jenkins. Typical output is as follows:

```
a=(1+2i),(3+4i),(5+6i),(7+8i)
z=(-0.120851+1.7188i),(-1.75701+0.207073i),(-0.322141-
   1.52588i)
a=(-9.959+8.467i),(-3.666-3.501i),(9.169+5.724i),(1.478-
   0.643i)
z=(-0.0832328+0.111289i),(0.762881+0.306844i),(-
   0.719835-0.803841i)
a=(-3.039-5.537i),(-4.295-1.856i),(-6.72+6.827i),(-
   0.039-9.509i)
z=(0.410517-0.484651i),(-1.60023-
   0.370736i),(0.604929+1.31012i)
a=(-7.005+1.942i),(-5.173-4.564i),(2.39+4.604i),(-6.098-
   9.847i)
z=(0.15504+0.936837i),(0.694713-0.680799i),(-1.36779-
   1.05119i)
a=(-9.708+2.382i),(7.421+8.716i),(9.718+9.895i),(-4.553-
   8.275i)
z=(0.517487+0.0547907i),(-1.0417-
   0.329409i),(1.03744+1.29836i)
a=(4.771+1.538i),(-8.131+9.912i),(-4.334-3.702i),(7.035-
   0.106i)
z=(0.650561+0.184551i),(-0.760413-0.427239i),(1.04699-
   2.13696i)
a=(-1.298-6.19i),(1.321+0.332i),(7.673-5.336i),(5.141-
   2.289i)
z=(-0.497039-0.170566i),(-0.36336+1.09371i),(0.954641-
   1.11679i)
a=(-1.748-3.132i),(-4.454-2.357i),(2.661+2.756i),(-
   9.964+2.859i)
z=(0.971482+0.806244i),(-0.0407683-0.989109i),(-
   2.10971+0.94695i)
a=(-1.277-0.259i),(-2.472-9.222i),(2.316-6.965i),(-
   7.811-8.158i)
z=(0.0362865+0.807379i),(-0.837316-1.44827i),(-2.46508-
   5.91829i)
a=(-9.712+0.105i),(-0.96-1.058i),(9.264-7.353i),(-2.555-
   6.196i)
z=(-0.0692952+0.47559i),(-1.13555+0.0243696i),(1.10719-
   0.609953i)
```

## Chapter 7. A/C Circuits

Alternating current circuits is a clever application of complex variables. We will restrict our discussion to constant-frequency circuits, thus greatly simplifying the mathematics. In this case, the response of the circuits is cyclical in time and the same for each cycle. We also presume the forcing (voltage or current source) to be sinusoidal. We know that the voltage and current must be some combination of sin() and cos(). By extension of Equation 1.5, we can represent voltage and current by some complex exponential, as in the following:

$$V(t) = ae^{i\omega t}$$
$$I(t) = be^{i\omega t} \qquad (7.1)$$

where $\omega$ is the frequency (radians per second). $A$ and $B$ are complex constants. Because voltage and current are real quantities, we can even combine these two into a single complex combination, choosing voltage as the real part and current the ideal part. From Ohm's Law we know that:[23]

$$V = IR \qquad (7.2)$$

The current and voltage in an inductor is related by what is often called Lenz's Law:[24]

$$V_L = -L\frac{dI}{dt} \qquad (7.3)$$

For a capacitor, we have a similar relationship:[25]

$$I_C = C\frac{dV}{dt} \qquad (7.4)$$

The corresponding integral form of the preceding relationship is also useful:

$$V_C = \frac{1}{C}\int I\,dt \qquad (7.5)$$

In these two equations we see complementary functions of voltage, current, and time. In the case of sin() and cos() we expect the temporal derivatives to yield a factor of $\omega$ and so we often see the products $\omega C$ and $\omega L$. The integral will likewise yield the product $1/\omega C$. We will first consider a series RLC circuit, as shown in the following figure.

---

[23] Georg Simon Ohm (1789–1854) German physicist and mathematician.
[24] Heinrich Friedrich Emil Lenz (1804–1865) Russian physicist of Baltic German ethnicity.
[25] This equation is often attributed to British scientist Michael Faraday (1791–1867).

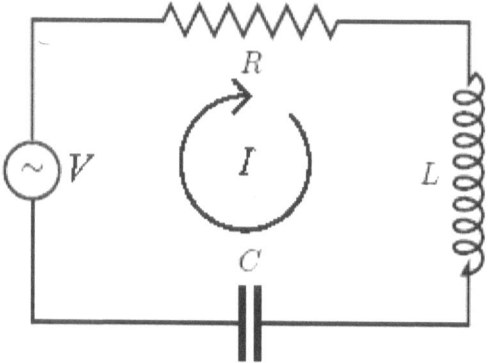

There is a single current that flows through the entire circuit and each of the components. This may be set to the following equation:

$$I = b\cos(\omega t) \tag{7.6}$$

The voltage across the resistor follows from Equation 7.2:

$$V_R = bR\cos(\omega t) \tag{7.7}$$

The voltage across the inductor follows from Equation 7.4:

$$V_L = b\omega L\sin(\omega t) \tag{7.8}$$

The voltage across the capacitor follows from Equation 7.6:

$$V_C = \frac{-b}{\omega C}\sin(\omega t) \tag{7.9}$$

These can be combined to form:

$$V = b\cos(\omega t) + b\left(\omega L - \frac{1}{\omega C}\right)\sin(\omega t) \tag{7.10}$$

When $\omega L = 1/\omega C$ the second term vanishes. The first term is the non-reactive part and the second term is called the reactive part of the solution (or response). The most commonly encountered frequency is 60 Hz (or 50 Hz). These equations have been implemented in an Excel® spreadsheet called circuit1.xls, which you will find in the folder examples\circuits. The results are shown in the following figure.

From Equation 7.3 we see that a Henry (the unit of measure for an inductor) is equal to a volt-second per amp. Noting that an ohm is equal to a volt per amp, a Henry is equal to an ohm-second.

From Equation 7.4 we see that a Farad (the unit of measure for a capacitor) is equal to an amp-second per volt. Noting that an ohm is equal to a volt per amp, a Farad is equal to a second per ohm. Also noting that an amp is equal to a coulomb per second, a Farad is equal to a coulomb per volt. A Farad is also equal to a second-squared per Henry.

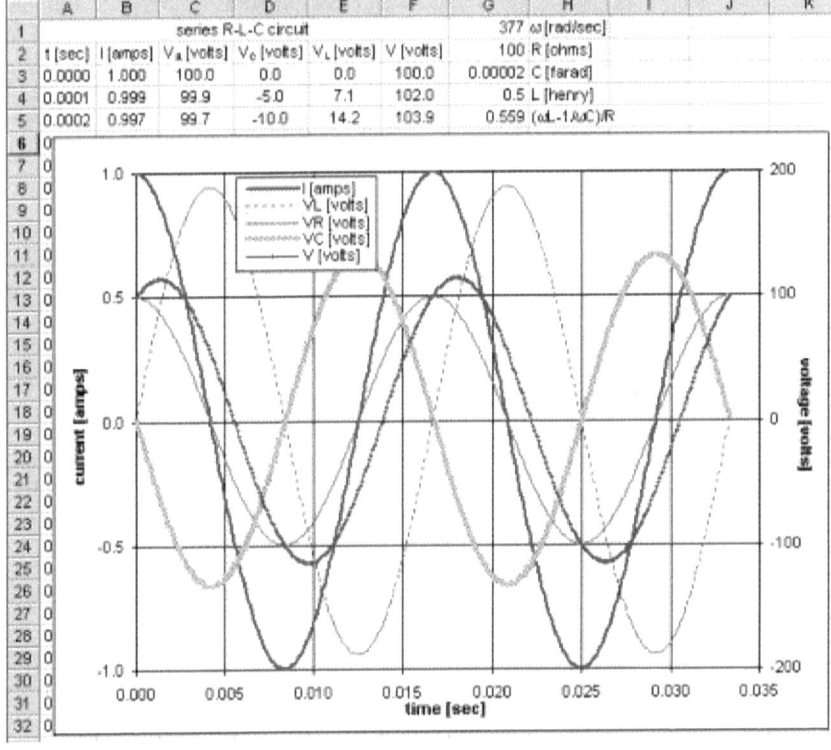

The results are shown above are for C=0.00002 Farad and L=0.5 Henry. When the inductance, capacitance, and frequency combine to result in a zero numerator, the phase angle between voltage and current disappears and the supply sees a non-reactive load. This occurs at 316.228 radians/second. The following figure shows the result.

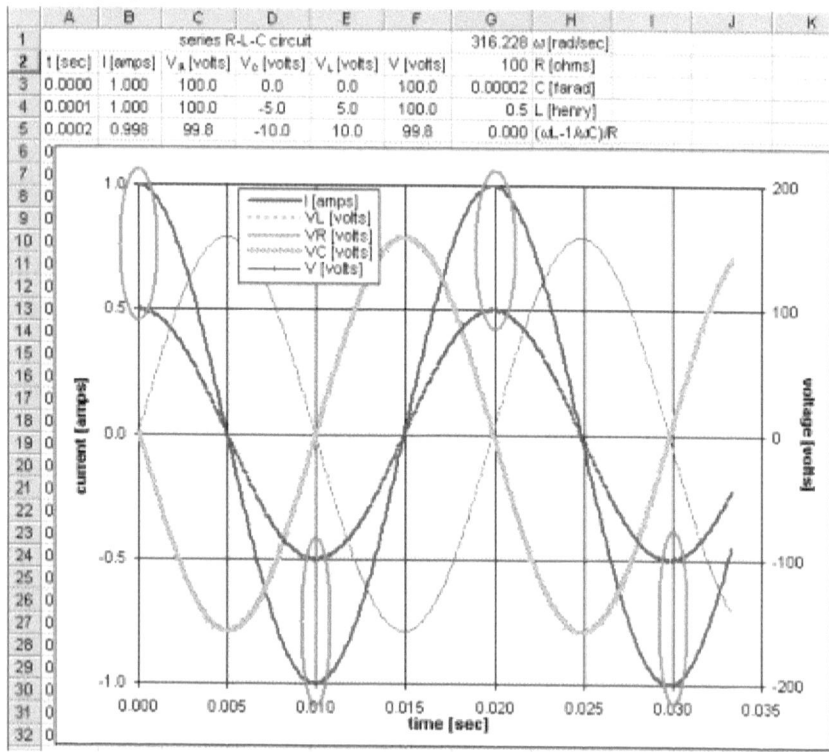

The current (solid blue) and voltage (violet +s) are in phase (i.e., the peaks and troughs coincide, as indicated by the magenta ellipses).

We can also solve this problem using complex variables via Euler's formula (Equation 1.5). Equation 7.7 becomes:

$$V_R = bR\cos(\omega t) = real\{b\,\text{Re}^{i\omega t}\} \tag{7.11}$$

Noting that $sin(x)=cos(x-\pi/2)$, Equation 7.8 becomes:

$$V_L = b\omega L\sin(\omega t) = real\left\{b\omega L e^{i\omega t - \frac{i\pi}{2}}\right\} \tag{7.12}$$

Noting that $-sin(x)=cos(x+\pi/2)$, Equation 7.9 becomes:

56

$$V_C = \frac{-b}{\omega C}\sin(\omega t) = real\left\{\frac{b}{\omega C}e^{i\omega t + \frac{i\pi}{2}}\right\} \qquad (7.13)$$

These can be combined to form:

$$V = real\left\{bRe^{i\omega t} + b\omega Le^{i\omega t - \frac{i\pi}{2}} + \frac{b}{\omega C}e^{i\omega t + \frac{i\pi}{2}}\right\} \qquad (7.14)$$

There is a little program (circuit1.cpp) in this same folder that contains these formulas (7.11-7.14) and shows that these are equivalent to the previous (7.7-7.10). The mnemonic "ELI the ICE man" is used to remember the phase between voltage (E) and current (I) in an inductor (L) and a capacitor (C). This is illustrated in the following figure.

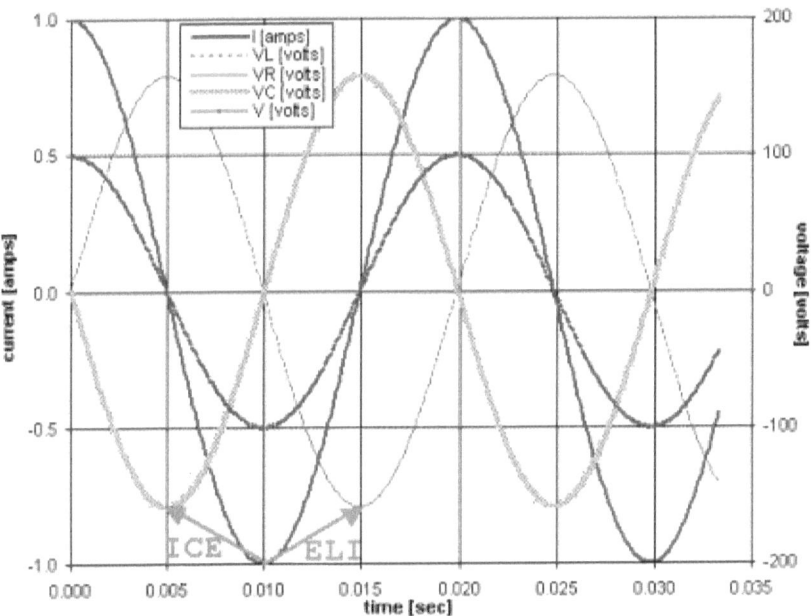

The peaks and troughs in the dotted brown voltage curve ($V_L$) *lead* (i.e., are "ahead" to the right in time) of the blue current curve (I). The peaks and troughs in the fuzzy green voltage curve ($V_C$) *lag* (i.e., are "behind" to the left in time) of the blue current curve (I). Noting that $e^{i\pi/2}=i$, Equation 7.14 becomes:

$$V = real\left\{b\left(R - i\omega L + \frac{i}{\omega C}\right)e^{i\omega t}\right\} \qquad (7.15)$$

Equation 7.6 can also be expressed as:

$$I = \text{real}\{be^{i\omega t}\} \qquad (7.16)$$

Comparing Equations 7.15 and 7.16, we can extract the term:

$$Z = R - i\omega L + \frac{i}{\omega C} \qquad (7.17)$$

We can then express Equation 7.2 as:

$$V = ZI \qquad (7.18)$$

This term Z is called impedance, as it's similar in function to R, only it is complex and may have an imaginary component. Equation 7.17 is linear, that is, the impedance of this series circuit is the sum of three terms: $R$, $-i\omega L$, and $i/\omega C$. These are the impedances of the resistor, inductor, and capacitor, respectively. The impedance, Z, can be represented by a magnitude, A, and angle, φ:

$$A^2 = R^2 + \left(\omega L - \frac{1}{\omega C}\right)^2 \qquad (7.19)$$

$$\tan \varphi = \frac{\omega L - \dfrac{1}{\omega C}}{R} \qquad (7.20)$$

If we use complex algebra, impedances in an A/C circuit (subject to the qualifications stated before) can be added just like resistors. This means that for the following parallel circuit, the impedance is given by:

$$\frac{1}{Z} = \frac{1}{R} - \frac{1}{i\omega L} + \frac{\omega C}{i} \qquad (7.21)$$

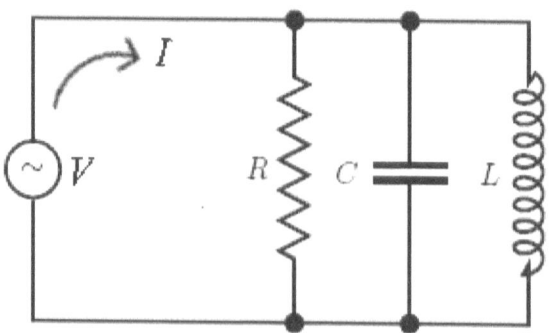

The code is in circuit2.cpp, an excerpt of which is listed below:

```
{
double t;
complex i,v,z;
```

```
z=1./(1./R-1./I/omega/L+omega*C/I);
for(t=0.;t<=1./30.;t+=0.0001)
   {
   i=b*exp(I*omega*t);
   v=z*b*exp(I*omega*t);
   printf("%1G,%1G,%1G\n",t,i.re,v.re);
   }
```

The output is graphed in the spreadsheet circuit2.xls and shown in the following figure:

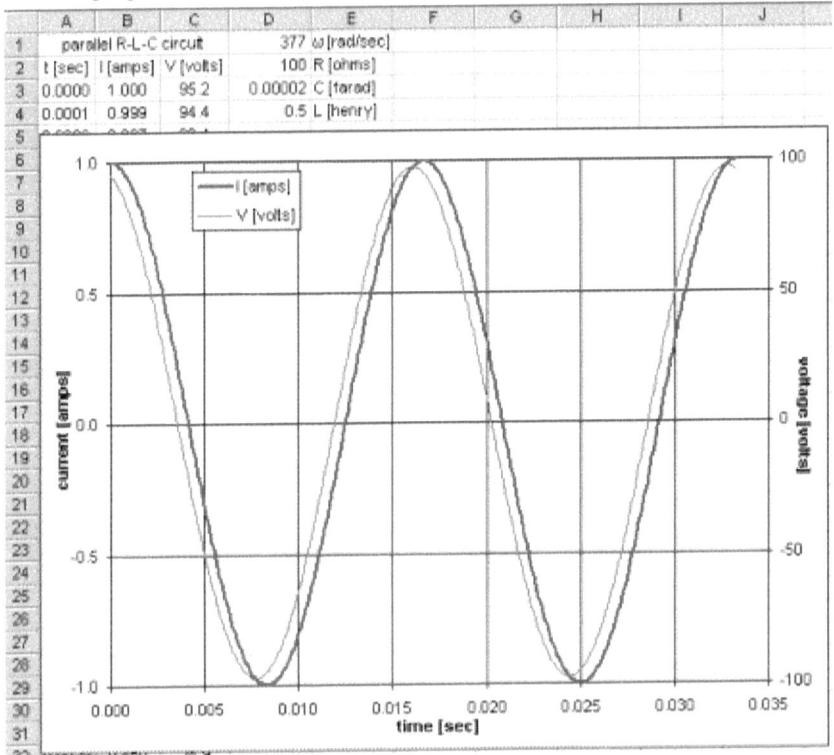

Again, the voltage and current curves lie on top of each other (i.e., the phase angle disappears) when $\omega^2 LC=1$.

## Chapter 8. Simultaneous Equations

Simultaneous equations involving complex variables are solved in the same way as real variables with a few modifications. For a general, non-iterative solution, Gauss elimination with row and column pivoting is the preferred method. Row pivoting is almost always necessary, as there is no reason to expect the equations to be eliminated in the same order as they are placed in the matrix. While it is difficult to prove analytically that column pivoting increases the accuracy of the solution, it is trivial to demonstrate the efficacy of this simple modification. Row and column pivoting simply means using the element of largest magnitude below and to the right of the current diagonal one in the process of reducing the matrix to upper triangular form.

In order to illustrate this process, we will use the common task of approximation. The target will be cosh(z) and the approximation will be a simple power series:

$$\cosh(z) \approx c_0 + c_1 z + c_2 z^2 + c_3 z^3 + \dots \qquad (8.1)$$

We will use 120 random values between (-2,-2i) and (+2,+2i). The coefficients, $c_0 \dots c_n$, are found by solving what is called the linear least-squares problem. This can be expressed in matrix form:

$$[c] = [A^T A]^{-1} [A^T B] \qquad (8.2)$$

The matrix *[A]* is:

$$[A] = \begin{bmatrix} 1 & z_1 & z_1^2 & z_1^3 \\ 1 & z_2 & z_2^2 & z_2^3 \\ 1 & z_3 & z_3^2 & z_3^3 \\ 1 & z_4 & z_4^2 & z_4^3 \end{bmatrix} \qquad (8.3)$$

The matrix *[B]* is:

$$[B] = \begin{bmatrix} \cosh(z_1) \\ \cosh(z_2) \\ \cosh(z_3) \\ \cosh(z_4) \end{bmatrix} \qquad (8.4)$$

The code to accomplish this task (llsq.cpp) can be found in folder examples\simultaneous. The associated code (cosh.cpp) is in this same folder.

The real (red) and imaginary (blue) components of cosh(z) are illustrated in the following figure:

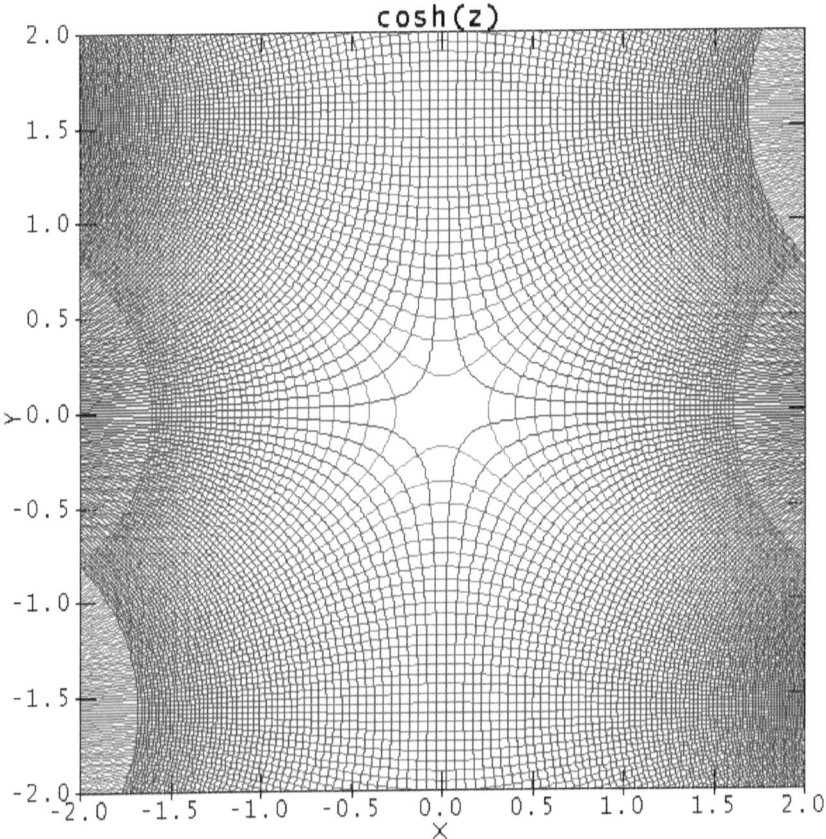

Abbreviated output of the program llsc.cpp is listed below:

```
calculating 120 values at random points
solving linear least-squares problem
error=0
c[9]=(+0.999927-0.000291482i),
     (+9.63084E-005+0.00060634i),
     (+0.499701-0.000157671i),
     (+0.000111215-8.02437E-05i),
     (+0.0416647-5.84014E-5i),
     (+5.67277E-006+2.18892E-5i),
     (+0.00136807-4.33395E-6i),
     (+3.0545E-006-2.53019E-6i),
     (+2.45044E-5-1.30648E-6i)
     random points     target value      regression
cosh(+0.463-1.959i)=(-0.420-0.444i)?(-0.417-0.440i)
```

```
cosh(+0.494+0.333i)=(+1.063+0.168i)?(+1.063+0.168i)
cosh(+1.721+1.165i)=(+1.139+2.486i)?(+1.139+2.487i)
cosh(-0.649+1.476i)=(+0.115-0.692i)?(+0.115-0.693i)
cosh(-1.542+0.956i)=(+1.410-1.822i)?(+1.409-1.821i)
cosh(-1.862-0.296i)=(+3.153+0.916i)?(+3.151+0.912i)
cosh(-1.177+1.276i)=(+0.516-1.405i)?(+0.516-1.405i)
cosh(-1.509-0.041i)=(+2.370+0.088i)?(+2.368+0.086i)
cosh(+1.940+0.995i)=(+1.934+2.858i)?(+1.935+2.860i)
cosh(-0.565-1.174i)=(+0.450+0.549i)?(+0.451+0.548i)
cosh(+0.601-1.617i)=(-0.055-0.637i)?(-0.054-0.635i)
cosh(-1.847+1.902i)=(-1.057-2.924i)?(-1.057-2.924i)
cosh(-1.621-1.708i)=(-0.359+2.407i)?(-0.360+2.408i)
cosh(+0.712-0.583i)=(+1.056-0.426i)?(+1.056-0.426i)
cosh(+1.891+1.714i)=(-0.484+3.204i)?(-0.485+3.204i)
cosh(-0.279-0.554i)=(+0.884+0.149i)?(+0.884+0.148i)
```

This approximation is just the first 9 terms, but appears to be adequate for this example. The number of points and terms is easily adjusted in the program, as is the target function.

## Chapter 9. Ordinary Differential Equations

ODEs involving complex variables can readily be solved using the same methods applied to real differential equations, the most common being Runge-Kutta. The fourth-order method is adequate for most problems. The ODE we will solve for an example is:[26]

$$\frac{d^3W}{dz^3} + W\frac{d^2W}{dz^2} = 0 \tag{9.1}$$

In general, the ODE must be rearranged to have the following form:

$$\frac{dy}{dx} = f(y(x), x) \tag{9.2}$$

The first step of the 4th order R-K method is defined:

$$k_1 = f(y(x), x) \tag{9.3}$$

The second step of the method is defined by:

$$k_2 = f\left(y(x) + k_1\frac{dx}{2}, x + \frac{dx}{2}\right) \tag{9.4}$$

The third step is:

$$k_3 = f\left(y(x) + k_2\frac{dx}{2}, x + \frac{dx}{2}\right) \tag{9.5}$$

The fourth step is:

$$k_4 = f(y(x) + k_3 dx, x + dx) \tag{9.6}$$

The last step in the process is:

$$y(x + dx) = y(x) + \left(\frac{k_1 + 2k_2 + 2k_3 + k_4}{6}\right)dx \tag{9.7}$$

As this is a third order ODE, we define an array of functions having three dimensions. The last (third) value dy[2]=f(y(x),x), the second value dy[1]=y[2], and the first value dy[0]=y[1]. The first, second, and third derivatives are integrated at each step of the process. In this way we solve the first, second, and third order ODE as the process progresses. The ODE is listed below:

```
void ODE(complex X,complex*Y,complex*dY)
    {
    dY[2]=-Y[0]*Y[2]/2.;
```

---

[26] This is the Blasius equation for the viscous boundary layer, a well-known problem of fluid mechanics.

```
dY[1]=Y[2];
dY[0]=Y[1];
}
```

The core function implementing the 4th order Runge-Kutta algorithm is listed below:

```
void RungeKutta4(void dYdX(complex,complex*,complex*),
    complex X,complex dX,complex*Y,complex*dY,int n)
{
int i,j;
double A[3]={0.5,0.5,1.};
double B[4]={1./6.,1./3.,1./3.,1./6.};
complex*V,*W;
V=(complex*)calloc(5*n,sizeof(complex));
W=V+n;
dYdX(X,Y,W);
for(i=1;i<4;i++)
   {
   for(j=0;j<n;j++)
      {
      dY[j]=A[i-1]*W[n*(i-1)+j];
      V[j]=Y[j]+dX*dY[j];
      }
   dYdX(X+dX*A[i-1],V,W+n*i);
   }
for(j=0;j<n;j++)
   {
   dY[j]=0.;
   for(i=0;i<4;i++)
      dY[j]+=B[i]*W[n*i+j];
   Y[j]+=dX*dY[j];
   }
free(V);
}
```

As this problem actually has two dimensions (the real and imaginary), we will begin at a known point and integrate outward, first to the right (increasing real component) and then upward (increasing imaginary component) until the domain is covered. The implementation is relatively simple:

```
void SolveODE()
   {
   int i,j;
   complex dX,dY[3],X,Y[3];

   /* initial conditions: X=(-3,-3) Y=1, dY/dX=0.1,
      d2Y/dX2=-0.01 */

   X.re=X.im=-3.;
   Y[0].re=Y[0].im=1.;
```

64

```
  Y[1].re=Y[1].im=0.1;
  Y[2].re=Y[2].im=-0.01;

/* step along x-axis (at lowest y-value) */

  Z[0]=X;
  W[0]=Y[0];
  dW[0]=Y[1];
  ddW[0]=Y[2];
  dX.re=6./(n-1);
  dX.im=0.;
  for(j=1;j<n;j++)
     {
     RungeKutta4(ODE,X,dX,Y,dY,3);
     X+=dX;
     Z[j]=X;
     W[j]=Y[0];
     dW[j]=Y[1];
     ddW[j]=Y[2];
     }

/* step along y-axis (one row of x-values at a time) */

  dX.re=0.;
  dX.im=6./(m-1);
  for(i=1;i<m;i++)
     {
     for(j=0;j<n;j++)
        {
        X=Z[n*(i-1)+j];
        Y[0]=W[n*(i-1)+j];
        Y[1]=dW[n*(i-1)+j];
        Y[2]=ddW[n*(i-1)+j];
        RungeKutta4(ODE,X,dX,Y,dY,3);
        Z[n*i+j]=X+dX;
        W[n*i+j]=Y[0];
        dW[n*i+j]=Y[1];
        ddW[n*i+j]=Y[2];
        }
     }
  }
```

The real (red) and imaginary (blue) components of the result are shown in the following figure:

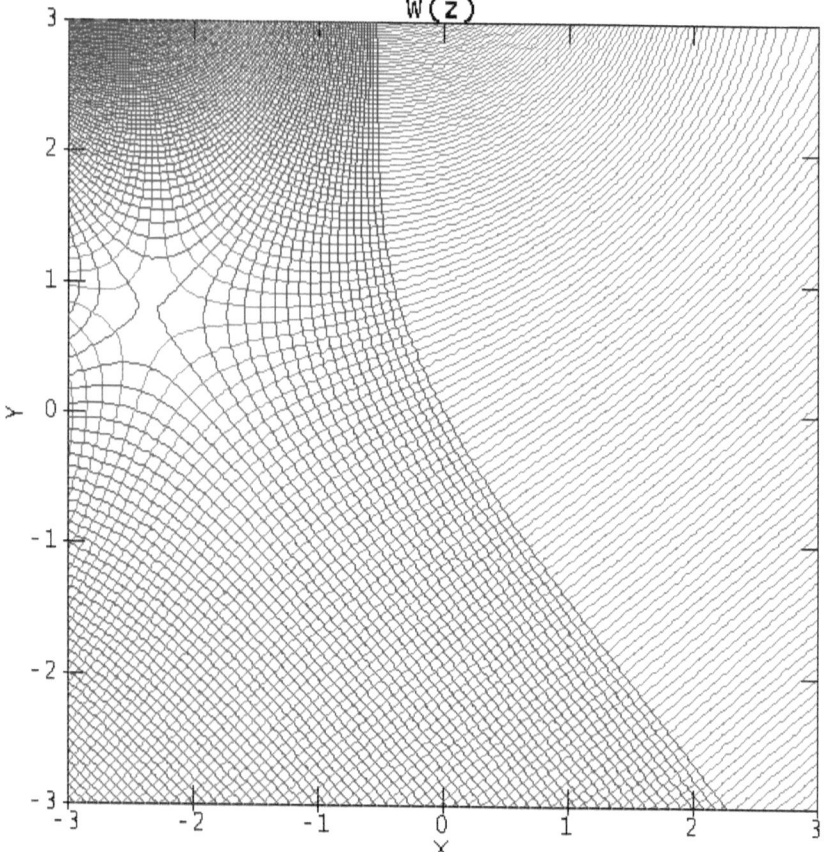

The real (red) and imaginary (blue) components of the first derivative of the result are shown in the following figure:

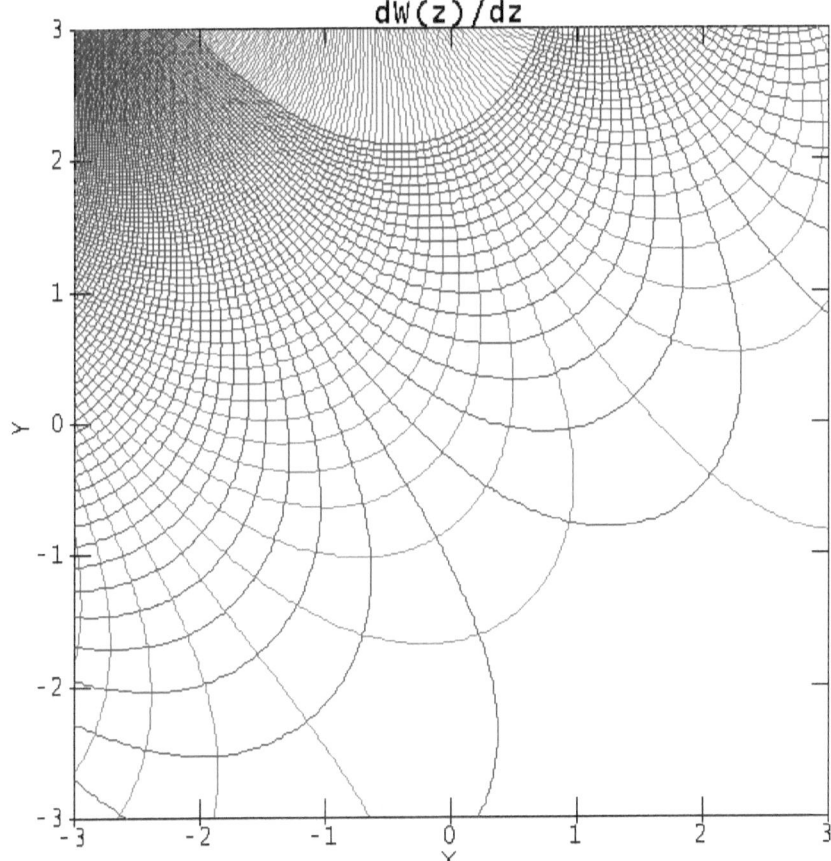

The real (red) and imaginary (blue) components of the second derivative of the result are shown in the following figure:

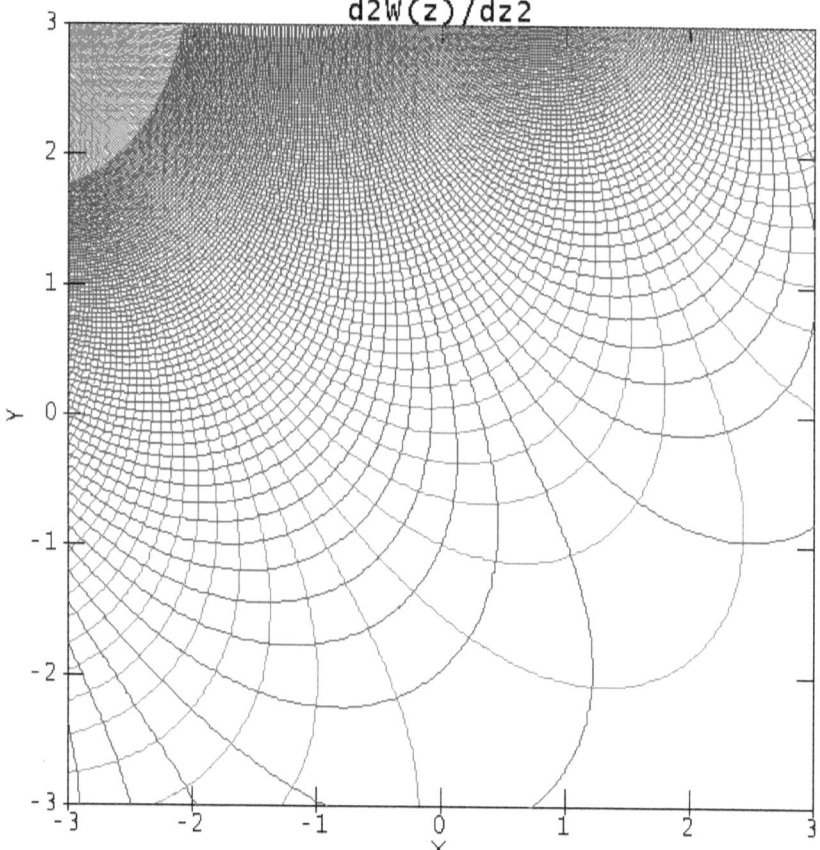

All of the associated files can be found in the folder \examples\Runge-Kutta.

## Appendix A: C++ Implementation

There are several Microsoft® include files for complex variables. These are a dreadful mess and inconsistent between one version of their compiler and the next. Complex variables were around long before Bill Gates was born, plus Walter Bright[27] invented the C++ compiler, so there's no excuse for this. I have provided a simple file that handles everything without all the fuss (complex.hpp in folder examples\include of the online archive). Below is an excerpt:

```
class complex
{
 public:
    friend double real (complex&);
    friend double imag (complex&);
    friend complex cos (complex&);
    friend complex sin (complex&);
    friend complex operator + (complex&,complex&);
    friend complex operator - (complex&,complex&);
    friend complex operator * (complex&,complex&);
    friend complex operator / (complex&,complex&);
 public:
    complex ();
    complex (double r,double i = 0);
    complex (complex &);
    double& real ();
    double& imag ();
    double re;
    double im;
    };
inline double real (complex& z) { return z.re; }
inline double imag (complex& z) { return z.im; }
inline double& complex::real () { return re; }
inline double& complex::imag () { return im; }
inline complex operator + (complex& lhs,complex& rhs)
  {
    return complex (lhs.re + rhs.re,lhs.im + rhs.im);
  }
```

All of the examples use this header. Many complex operations are performed just like scalar ones. The following is a simple example:

```
int main(int argc,char**argv,char**envp)
 {
   int i,j,n=10,m=10;
   double u,v,x,y;
```

---

[27] Walter Bright wrote Zortech®, the first single-pass C++ compiler. This was sold to Symantec®, who discontinued it because Microsoft® drove them out of the market. The same remarkable tool can now be found under the name Digital Mars.
https://www.digitalmars.com/

```
complex w,z;
for(i=0;i<n;i++)
   {
   u=-6.+i*7./(n-1);
   w.re=u;
   for(j=0;j<m;j++)
      {
      v=-9.+j*18./(m-1);
      w.im=v;
      z=exp(w);
      x=z.re;
      y=z.im;
      printf("%lG,%lG,%lG\n",x,y,v);
      }
   }
return(0);
}
```

## C++ Compilers

There are several C++ compilers available. I will mention only two: Digital Mars® and Microsoft®. Both are free. Digital Mars® is great, but doesn't provide 64-bit at this time, which is necessary to create Excel® AddIns for 64-bit versions of Microsoft® Office®. The Microsoft® compiler comes with the extremely annoying Visual Studio® IDDE that endlessly nags and defaults to every possible wrong setting (such as Unicode) and leaves hidden files and folders about like pig droppings.

You can avoid all of this hassle. Simply download and install the W7 SDK and DDK. Microsoft® has discontinued these, but they can be found elsewhere on the Web. Combine all of the folders (bin, include, and lib) to include the machine you intend to run it on (x86 or x64) and the machine you want to build for (x86 or x64). There are four combinations, although only two are essential. You can do the same with the W8 and W10 SDK and DDK packages; however, these will not run on every version of Windows® without patching a few dlls with a binary editor. The W10 package also requires some elements of Visual Studio®, so you would have to put up with the fussing either way.

## Appendix B. Common Complex Functions

The aforementioned include file (complex.hpp) also provides the common transcendental functions. Recall the Taylor series expansion for cos (Equation 1.6). If the argument includes $i$ ($\sqrt{-1}$), the alternating signs change with even powers due to $i^{2n}$. The Taylor series expansions for sinh and cosh are:

$$\sinh x = x + \frac{x^3}{3!} + \frac{x^5}{5!} + \ldots$$
$$\cosh x = 1 + \frac{x^2}{2!} + \frac{x^4}{4!} + \ldots \quad \text{(B.1)}$$

If follows then that the sin and cos of complex numbers are:

$$\sin(x+iy) = \sin x \cosh y + i \cos x \sinh y$$
$$\cos(x+iy) = \cos x \cosh y - i \sin x \sinh y \quad \text{(B.2)}$$

Notice that the preceding equations are very similar to the angle sum formulas: sin(a+b), cos(a+b). The tangent can be derived from B.2:

$$\tan(x+iy) = \frac{\sin 2x}{\cos 2x + \cosh 2y} + i \frac{\sinh 2y}{\cos 2x + \cosh 2y} \quad \text{(B.3)}$$

The hyperbolic sin and cos follow from B.1 and B.2:

$$\sinh(x+iy) = \sinh x \cos y + i \cosh x \sin y$$
$$\cosh(x+iy) = \cosh x \cos y - i \sinh x \sin y \quad \text{(B.4)}$$

And the hyperbolic tangent:

$$\tanh(x+iy) = \frac{\sinh 2x}{\cosh 2x + \cos 2y} + i \frac{\sin 2y}{\cosh 2x + \cos 2y} \quad \text{(B.5)}$$

The gamma function is also quite useful, as it appears in many analytical solutions. The definition is:

$$\Gamma(z) = \int_0^\infty t^{z-1} e^{-t} dt \quad \text{(B.6)}$$

There is no closed form solution and the Taylor series beginning at zero (or 1) is not practical; however, Sterling's expansion for large values converges quite well.

$$\Gamma(z) = e^{-z} z^{z-\frac{1}{2}} \sqrt{2\pi} \left( 1 + \frac{1}{12z} + \frac{1}{288z^2} - \frac{139}{51840z^3} + \ldots \right) \quad \text{(B.7)}$$

We use this along with the recursion relationship to evaluate the gamma function.

$$\Gamma(z+1) = z\Gamma(z) \tag{B.8}$$

We first run the value of $z$ up using B.8 until the magnitude is >100, then use B.7, then run it back down using B.8. The first and third steps can be combined by keeping track of both in one while() loop. The Riemann surface of the gamma function is shown in the following figure:

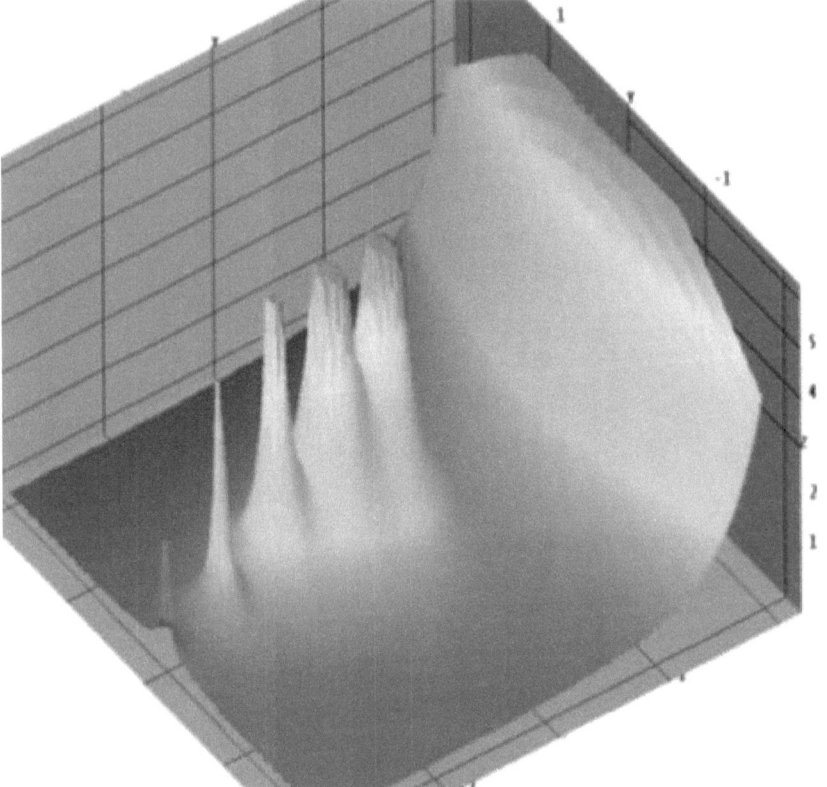

The poles at zero and the negative real integers have been clipped at z=6. The code to produce the surface is in examples\Riemann surfaces\rsurf.cpp

## Appendix C: Lambert's W

The Lambert function $W(z)$ is defined by the following relationship:

$$z = W(ze^z) \tag{C.1}$$

The Lambert function is included in complex.hpp and used in several figures and calculations throughout this text. The real and imaginary parts of the Lambert-W function are shown in the following two figures.

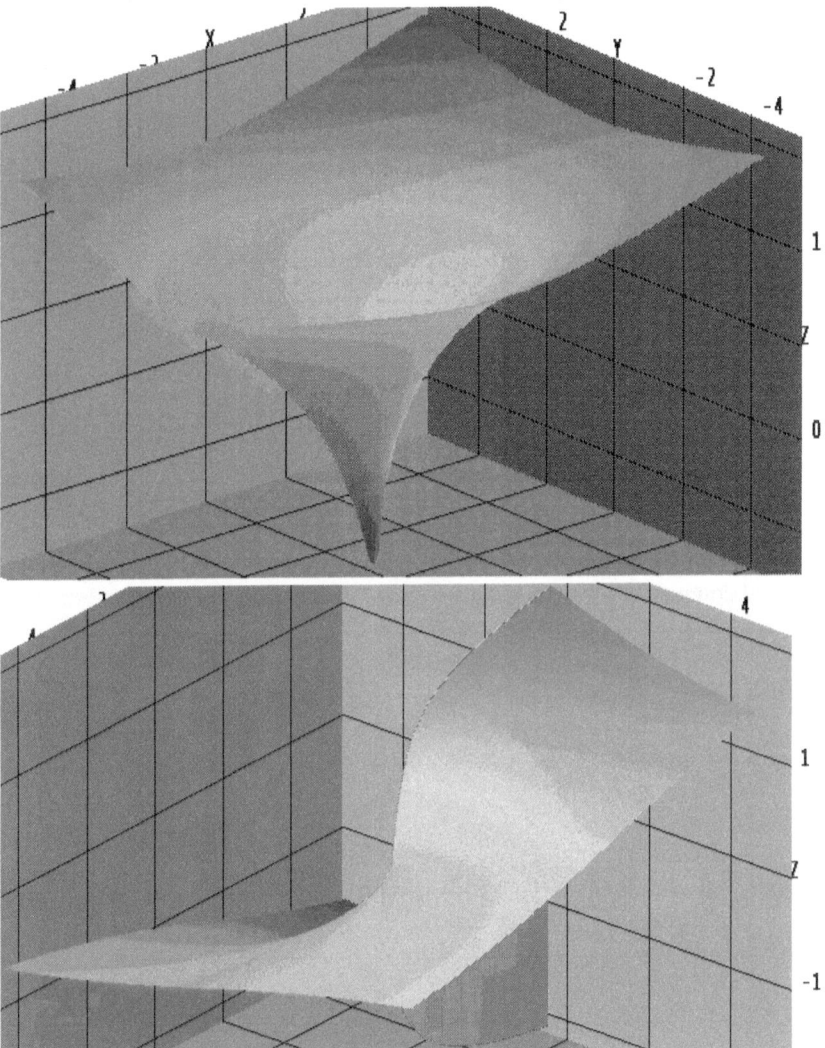

## Appendix D. Error Function

The Gauss error function is defined by the following integral. This function and its complement appear in the solution of a variety of problems.

$$erf(x) = \frac{2}{\sqrt{\pi}} \int_0^x e^{-t^2} dt \tag{D.1}$$

For real arguments, the relationship is quite simple:

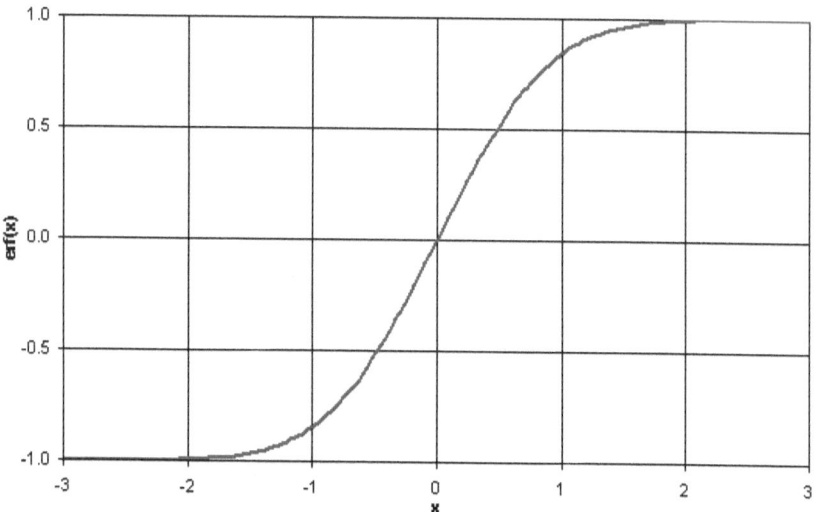

For complex arguments, the values are more complicated—even oscillatory. The real component for seval values is shown in this next figure:

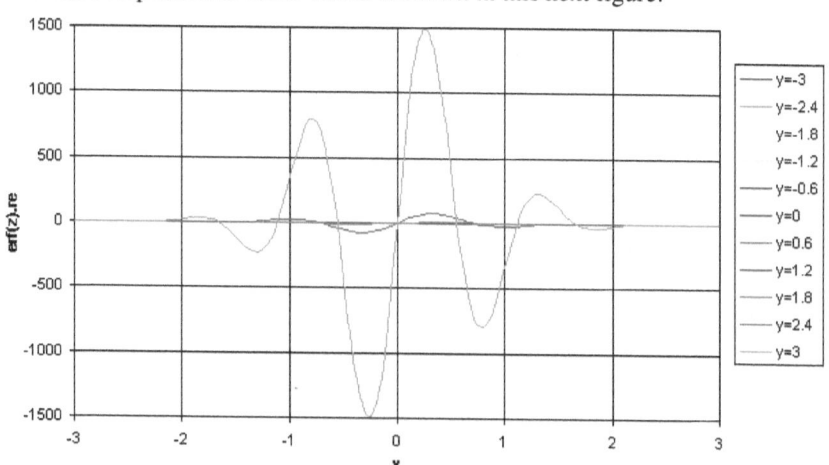

The imaginary component for the same values is shown in this next figure:

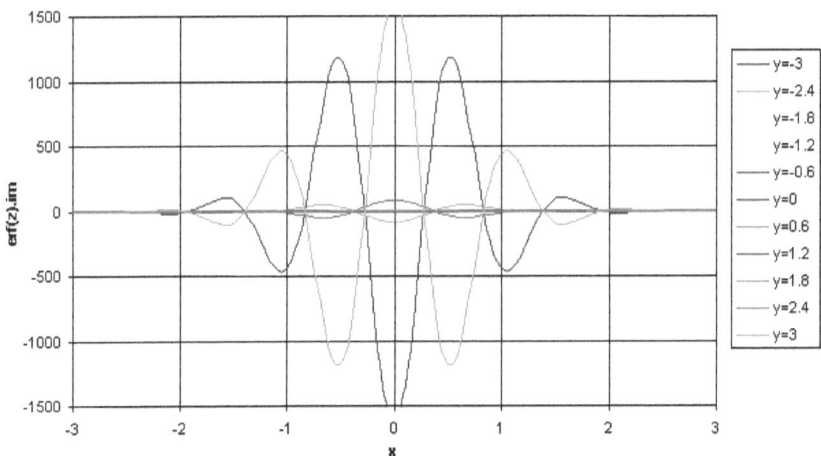

Three approximations described by Abramowitz and Stegun[28] are used to calculate the complex error function: Taylor series, rational polynomials, and an asymptotic expansion. The code (erf.cpp) and spreadsheet (erf.xls) can be found in the folder \examples\erf.

---

[28] Perhaps the most useful and complete reference on mathematical functions ever published is the *Handbook of Mathematical Functions* by Abramowitz and Stegun. It was first published by the National Bureau of Standards as Technical Monograph No. 55. The entire text is available free on-line at several locations.

Contours of the real (red) and imaginary (blue) components of the Gauss error function are shown in this next figure:

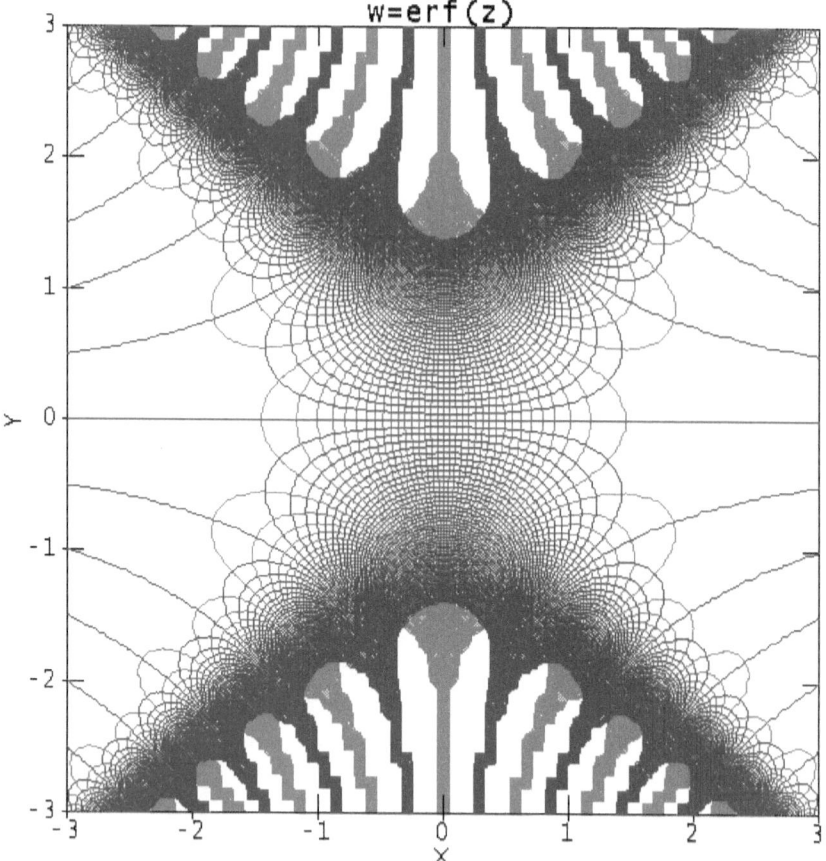

## Appendix E: Airy Function

The Airy function is a solution to the following ordinary differential equation. There are actually two forms (Ai and Bi), which show up in various complex differential equations and associated applications.

$$\frac{d^2 y}{dx^2} = xy \qquad (E.1)$$

There are several FORTRAN source codes available on the Web that purport to provide the first and second Airy functions for complex arguments. After downloading and compiling these with Digital Equipment Corporation's FORTRAN 90 Version 6.1, these functions produced nothing like the Airy functions as illustrated in both the Wikipedia and Wolfram Research web sites, which are consistent with each other. This failure may arise from some specialized constants and functions unique to some specific compiler, for example, the CRAY. This is an unacceptable situation. All code should be written to produce the same result, regardless of the compiler used, if not for every feature, certainly for mathematical functions.

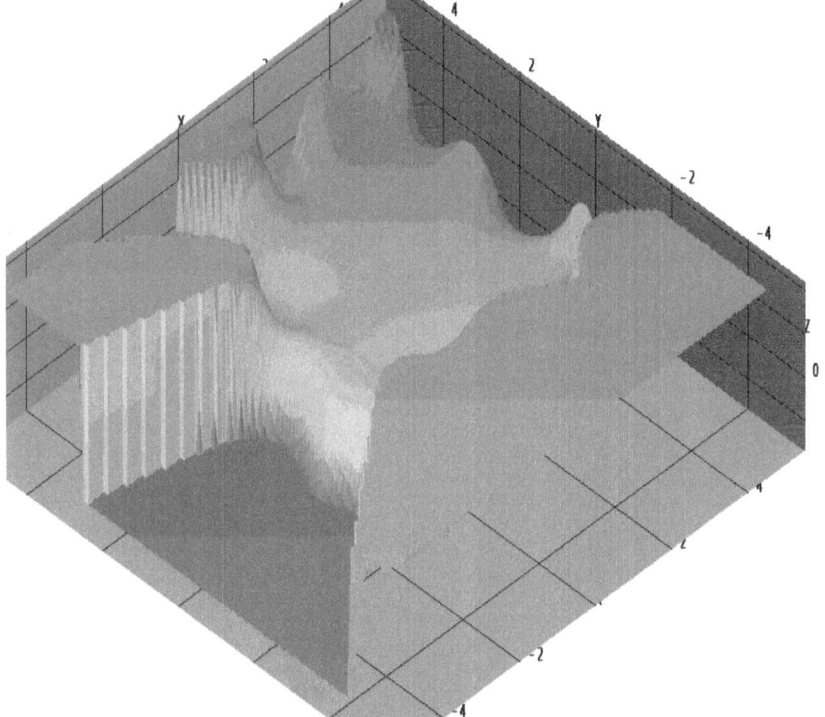

Maple® is the gold standard for mathematics, so I generated Ai and Bi for 10,000 random values and saved these in two files (AiryAi.csv and AiryBi.csv), which can be found in the folder examples\Airy, along with the script (Airy.mws). Tenth-order (66-term) regressions for each component of each function may be found in spreadsheet Airy.xls, along with a comparison of regression accuracy. The real component of the first kind, Ai, is shown in the preceding figure. The imaginary component is shown in this next figure:

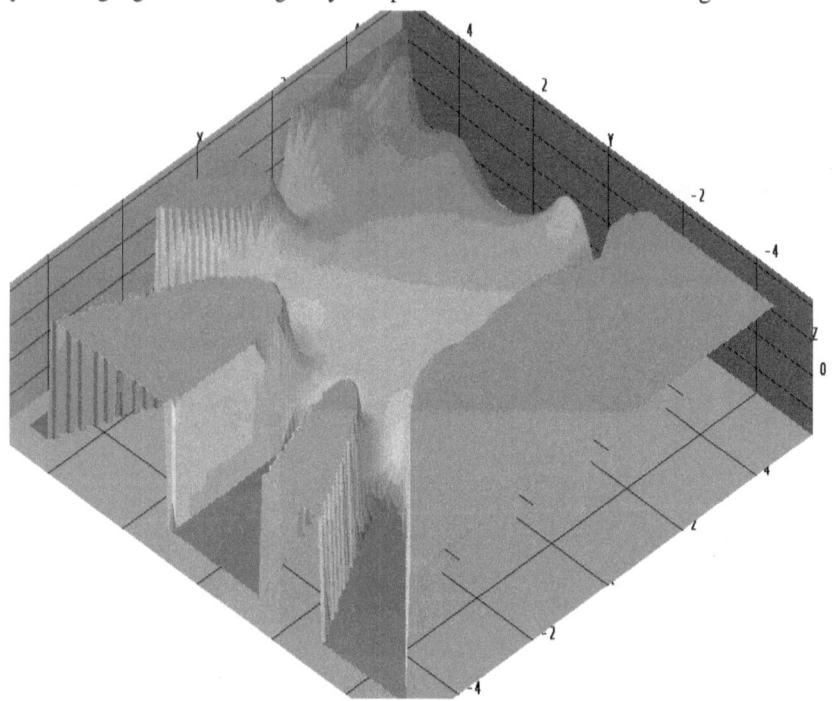

The real component of the second kind (Bi) is shown in the following figure:

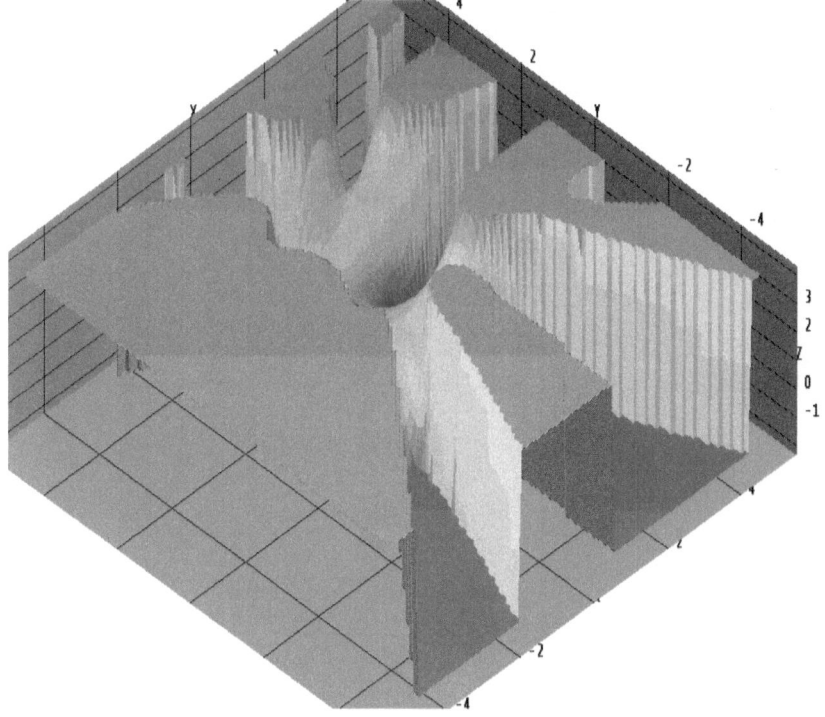

The imaginary component of the second kind (Bi) is shown in the following figure:

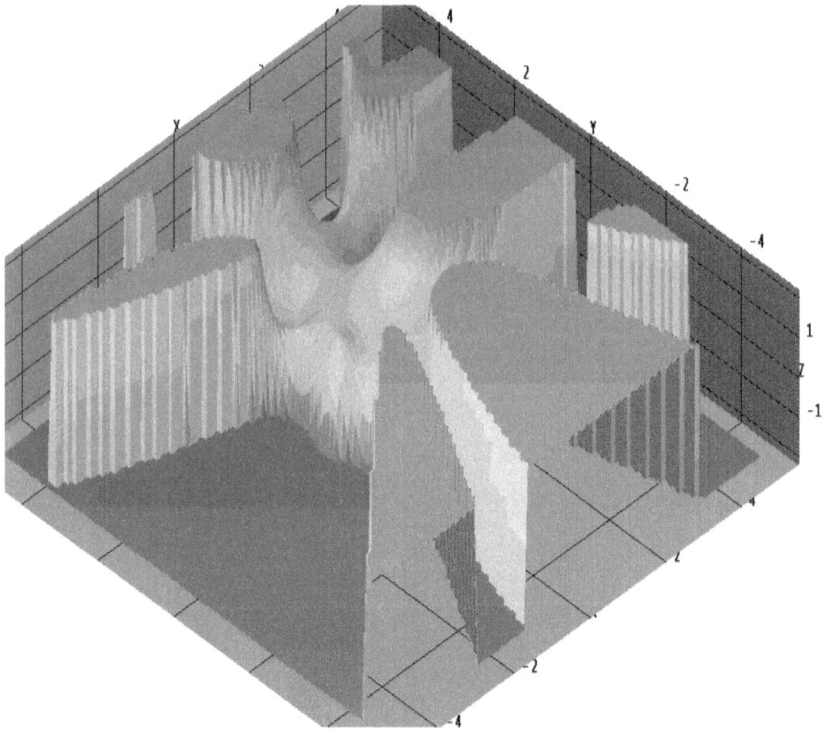

Contours of the real (red) and imaginary (blue) components of the first Airy function (Ai) are shown in this next figure:

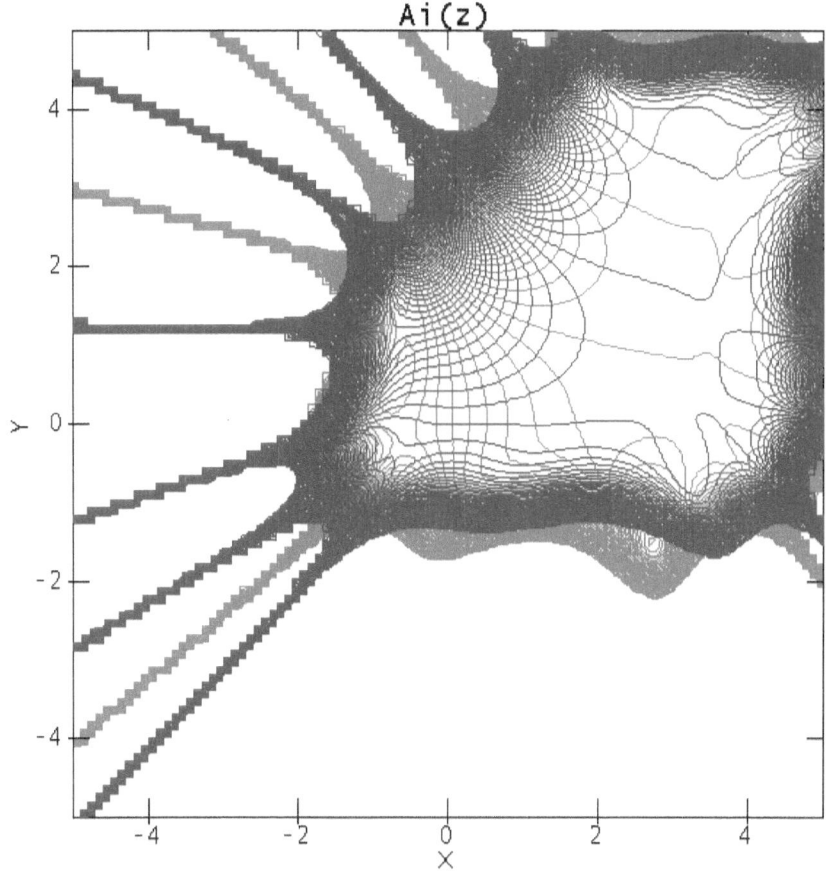

Contours of the real (red) and imaginary (blue) components of the second Airy function (Bi) are shown in this next figure:

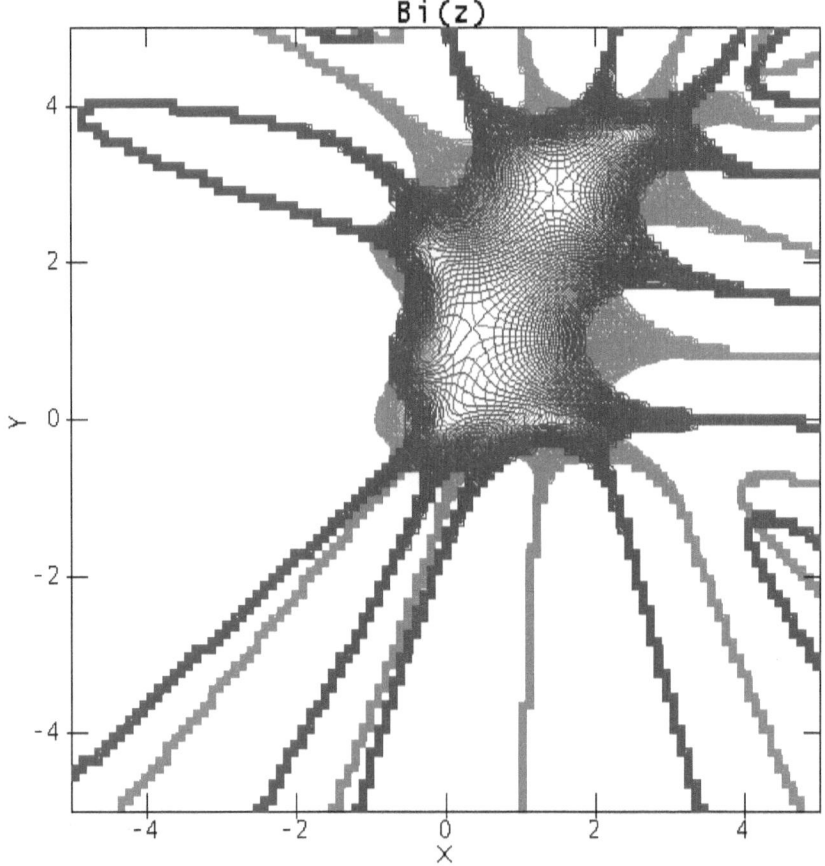

## Appendix F. Bessel Functions

Bessel functions of the first kind are solutions to the following second-order differential equation:

$$x^2 \frac{d^2 y}{dx^2} + x \frac{dy}{dx} + (x^2 - n^2) = 0 \qquad (F.1)$$

This differential equation arises in several fields, originally astrophysics.[29] It was first posed and solved by Bernoulli.[30] The solutions can be expressed as an infinite series:

$$J_n(x) = \sum_{m=0}^{\infty} \frac{(-1)^m}{m!\Gamma(m+n+1)} \left(\frac{x}{2}\right)^{2m+n} \qquad (F.2)$$

The solutions can also be expressed as an integral:

$$J_n(x) = \frac{1}{\pi} \int_0^{\pi} \cos(nt - x\sin t)\, dt \qquad (F.3)$$

This is equivalent to the complex form via Euler's identity:

$$J_n(x) = \frac{1}{2\pi} \int_{-\pi}^{\pi} e^{i(x\sin t - nt)}\, dt \qquad (F.4)$$

These relationships are valid even for complex arguments, which leads to modified Bessel functions of the first kind:

$$I_n(x) = i^{-n} J_n(ix) = \sum_{m=0}^{\infty} \frac{1}{m!\Gamma(m+n+1)} \left(\frac{x}{2}\right)^{2m+n} \qquad (F.5)$$

These modified Bessel functions of the first kind solve the following slightly different second-order differential equation:

$$x^2 \frac{d^2 y}{dx^2} + x \frac{dy}{dx} + (x^2 + n^2) = 0 \qquad (F.6)$$

---

[29] Friedrich Wilhelm Bessel (1784–1846) German astronomer, mathematician, and physicist.
[30] Daniel Bernoulli (1700–1782) Swiss mathematician and physicist, famous for his applications of mathematics to fluid mechanics and for his work in probability and statistics.

Both types are available in Excel® for real arguments (see Bessel.xls in folder \examples\Bessel). Orders zero through six of the unmodified form (BESSELJ) are shown in this next figure:

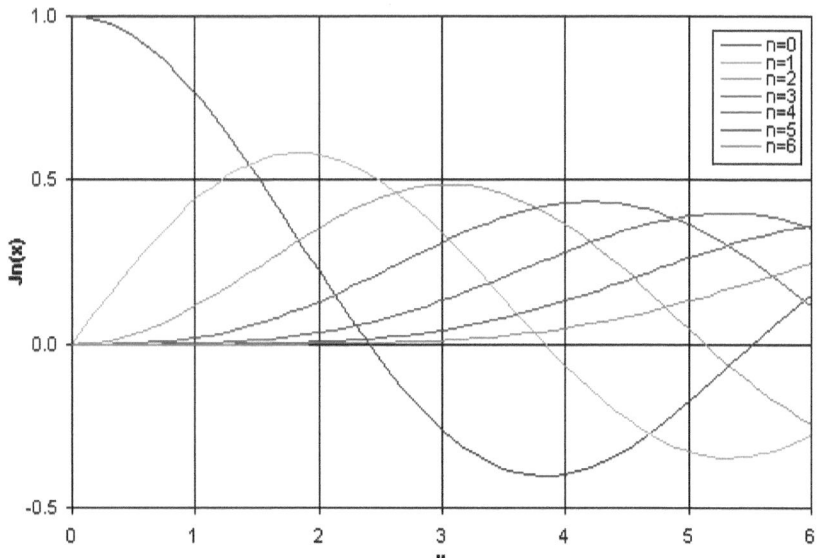

These exhibit oscillatory behavior. The modified type (BESSELI) can also be found in this same spreadsheet:

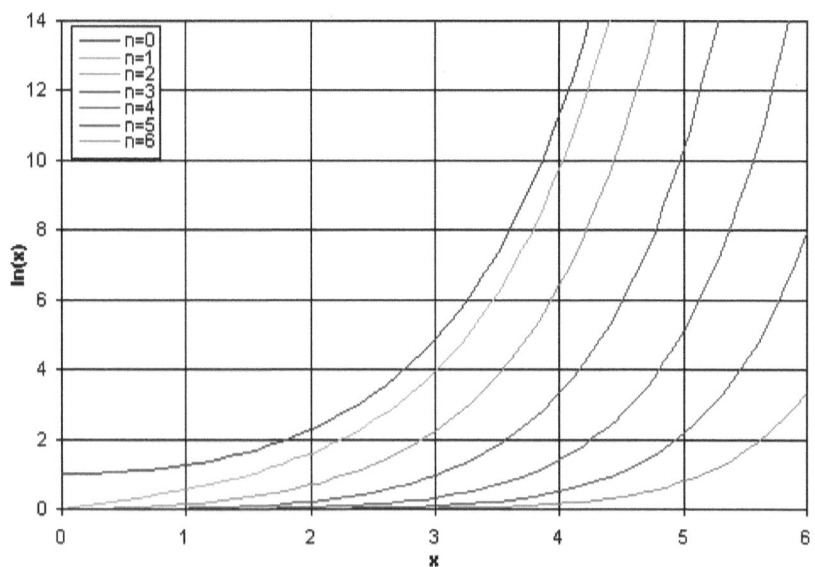

The infinite series is only practical for a limited range of values. There are several other approximations provided by Abramowitz and Stegun. These may be found in besselj.c and besseli.c in this same folder. These functions work for real and complex arguments. The real (red) and complex (blue) components of the unmodified Bessel function of the first type of zero order is shown in this next figure. Orders one and two are shown in the subsequent two.

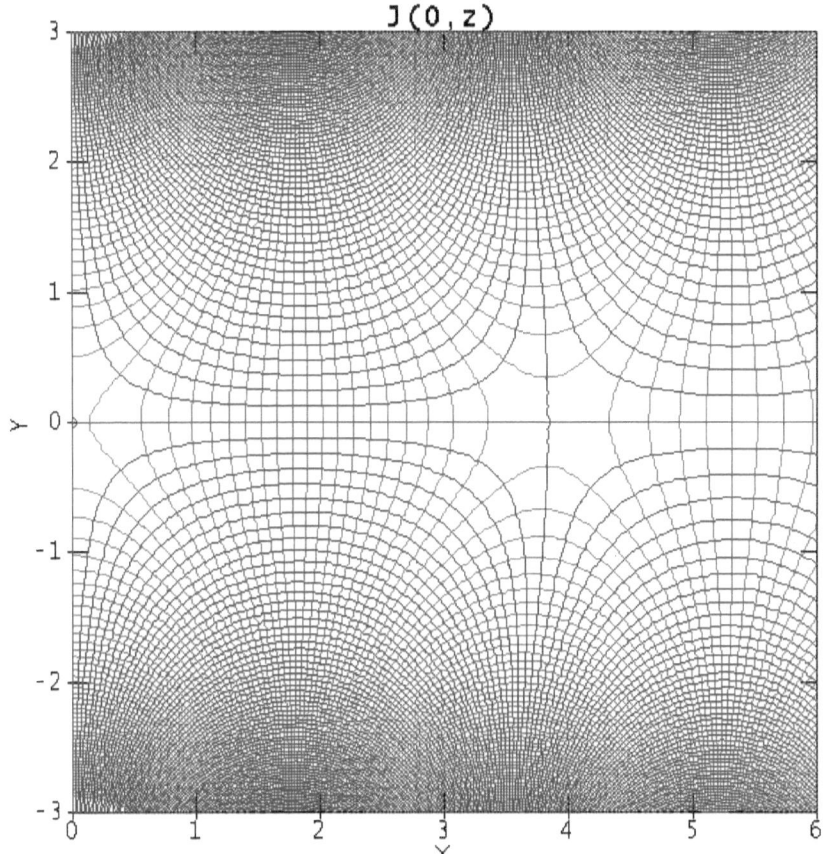

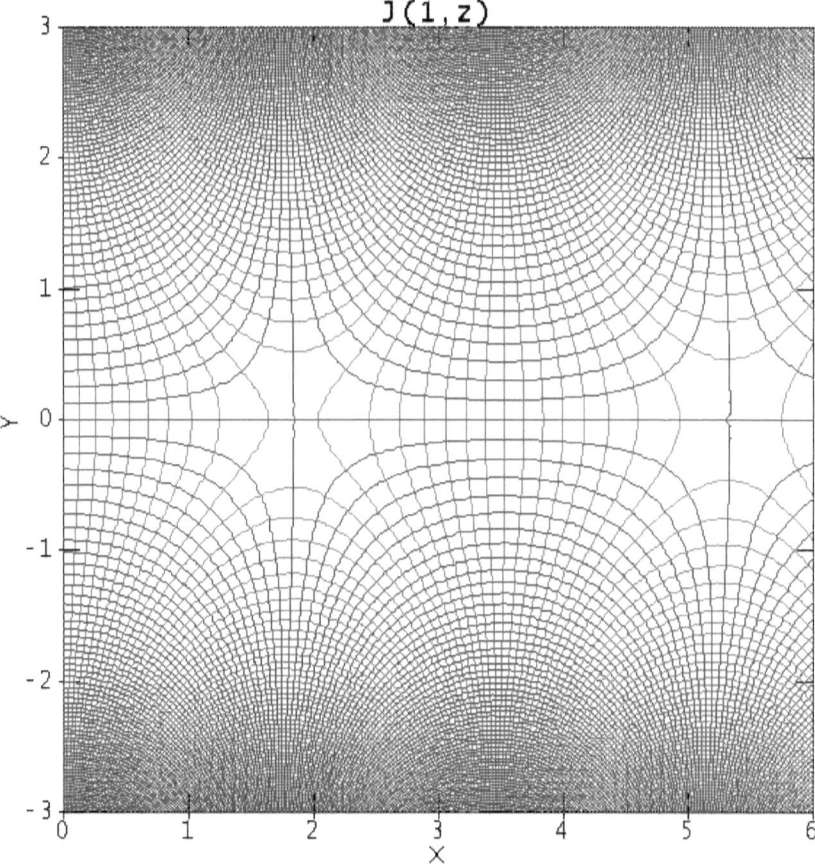

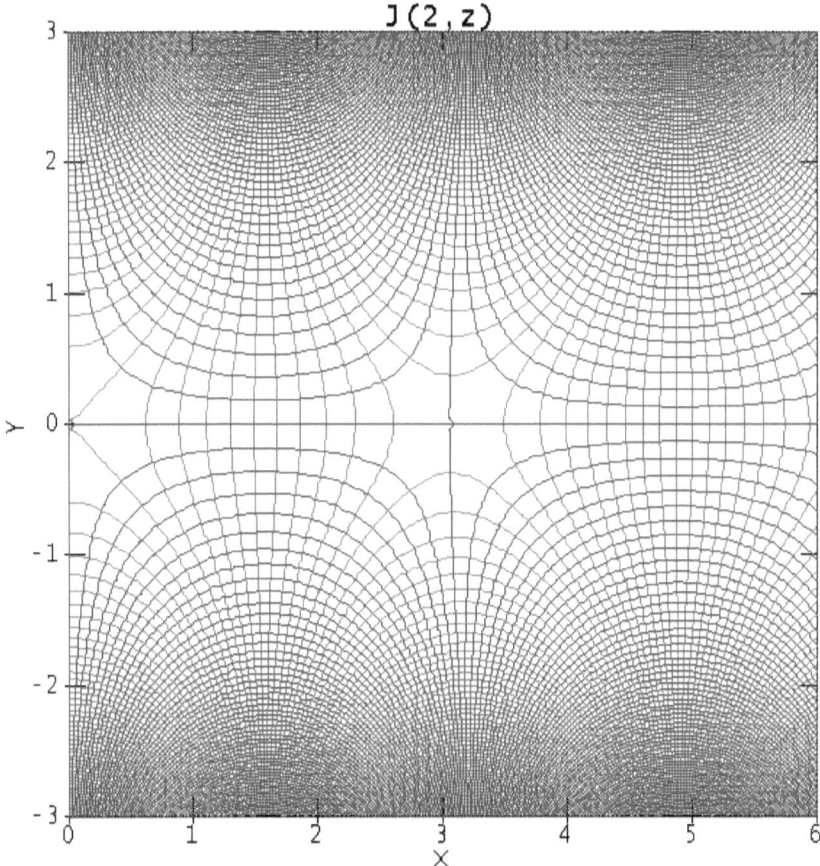

Orders zero, one, and two of the modified Bessel function of the first kind for complex arguments are shown in the next three figures:

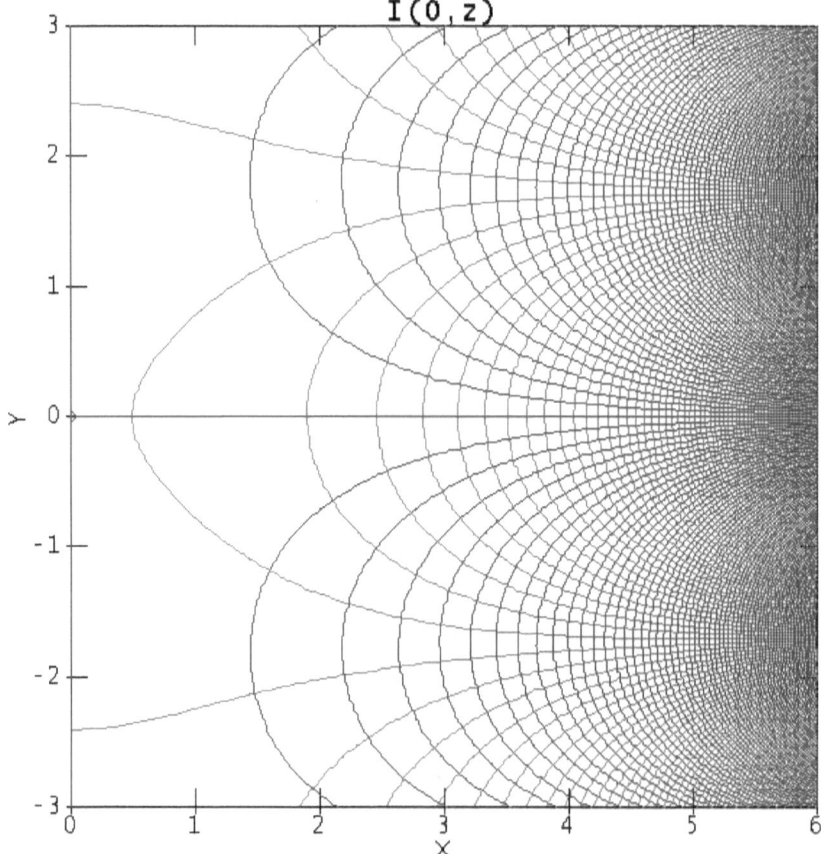

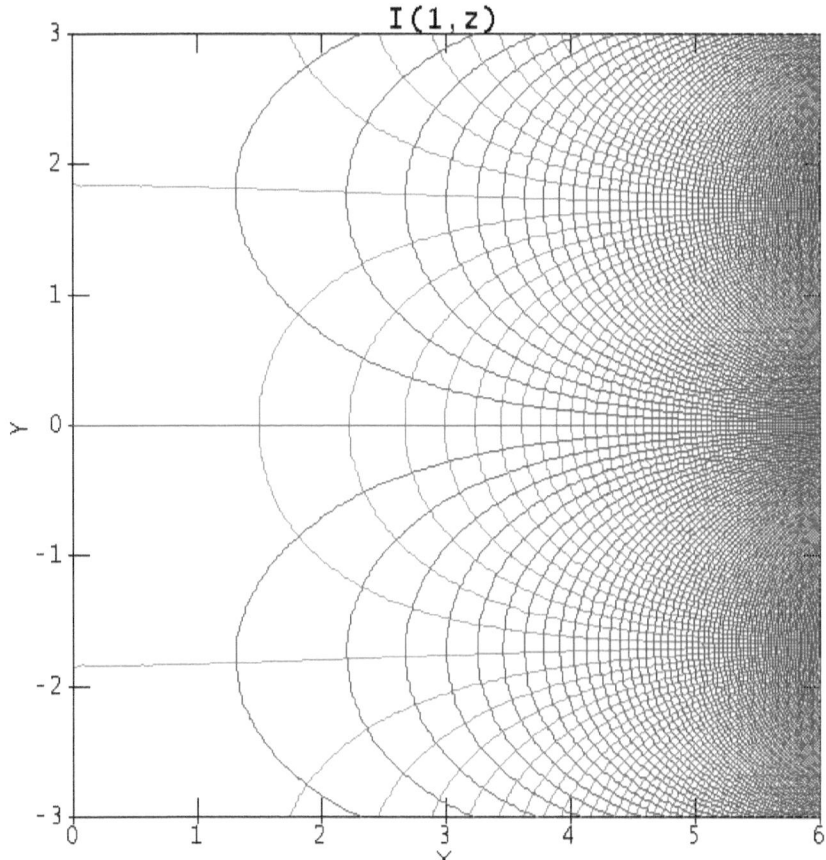

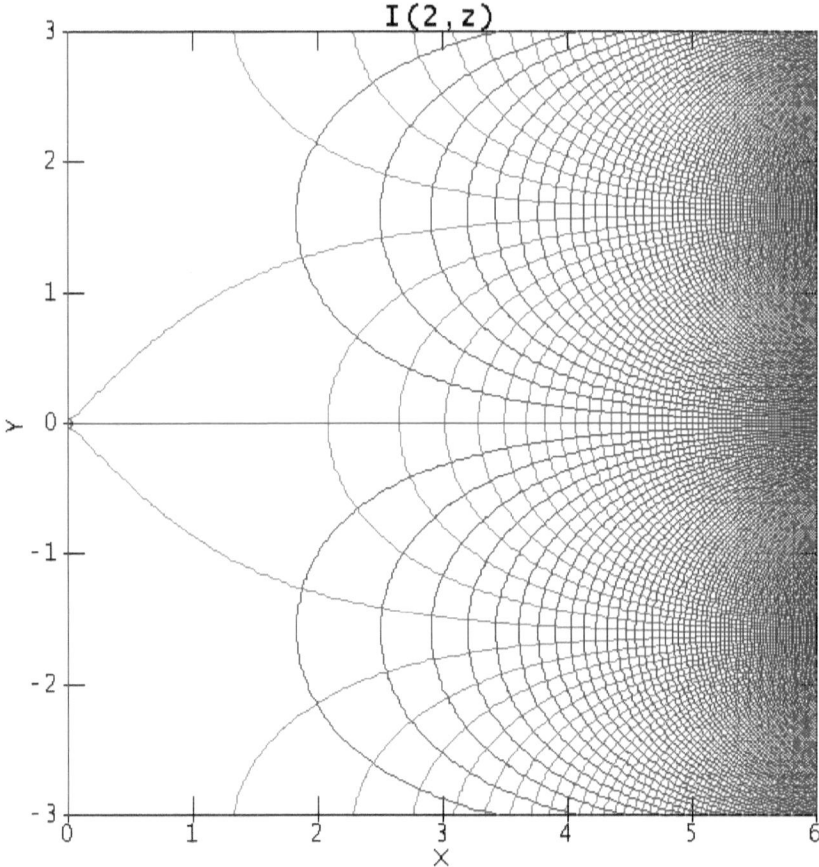

## Appendix G. Weierstrass Elliptic Functions

Weierstrass elliptic functions arise in the solution of waves. In particular, these functions named after Karl Weierstrass[31] are solutions to the Korteweg–de Vries equation,[32] which is similar to Bessel's equation and also second order.

$$\frac{d^2y}{dx^2} - (3y^2 - cy) = 0 \qquad (G.1)$$

The common sub-types are lemniscatic (following the arc of a lemniscate of Bernoulli) and equianharmonic (regular, non-harmonic). Both are approximated using the formulas provided by Abramowitz and Stegun. The code (weierstrass.cpp) can be found in folder examples\Weierstrass. The real (red) and imaginary (blue) components of the lemniscatic form are shown below:

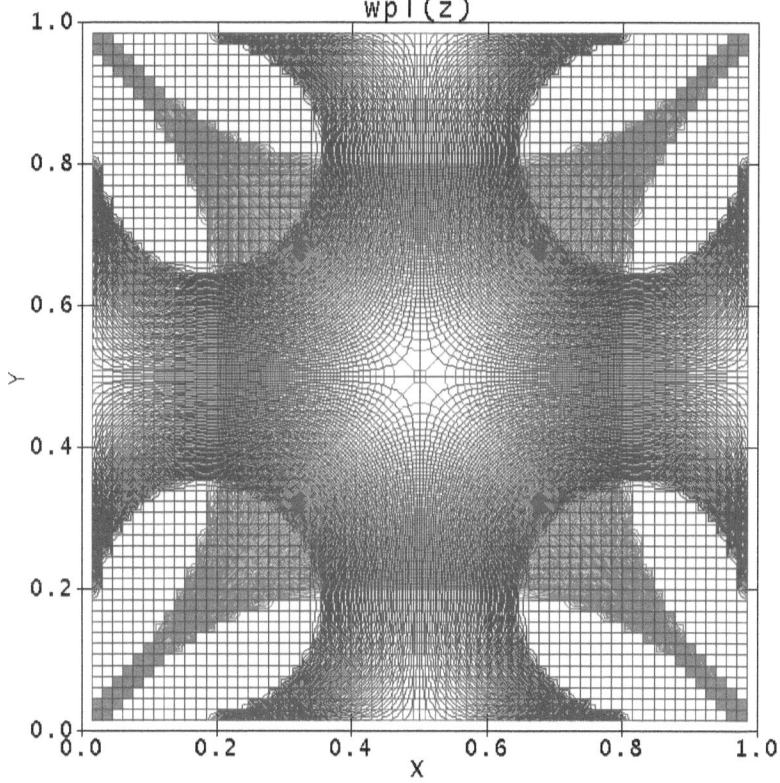

---

[31] Karl Theodor Wilhelm Weierstrass (1815–1897) German mathematician, physicist, and botanist.
[32] Shallow water wave equation named after Dutch mathematicians Diederik Johannes Korteweg (1848–1941) and Gustav de Vries (1866–1934).

The real (red) and imaginary (blue) components of the equianharmonic form are shown in the following figure:

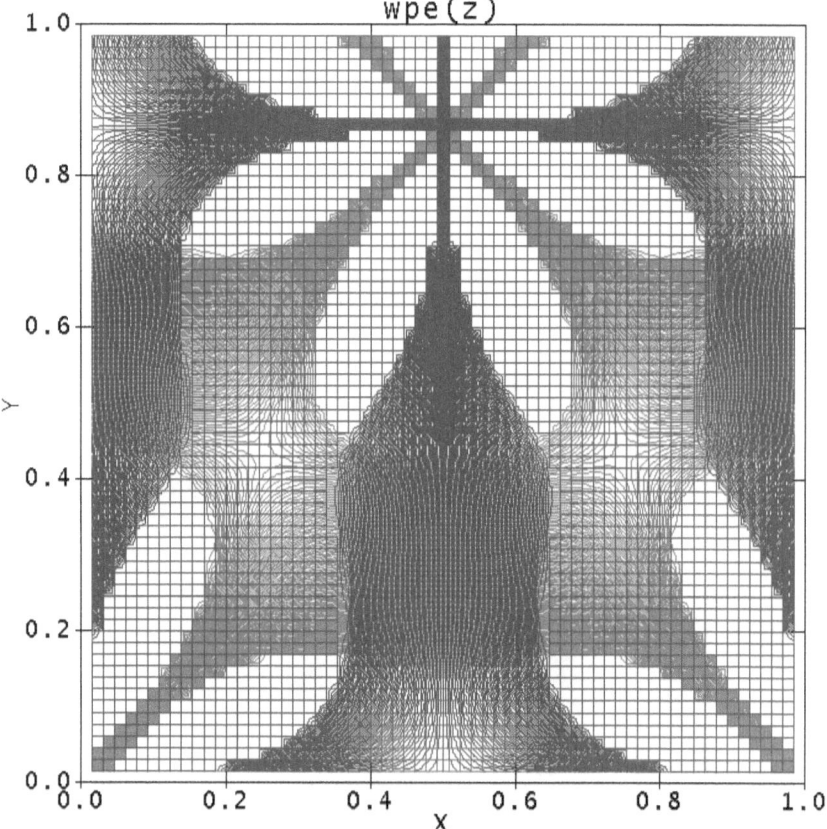

Even though the Weierstrass functions are uncommon, they produce such remarkable plots that they are often introduced for artistic reasons.

## Appendix H. Crazy Circular Plots

If you spend much time browsing the Web for complex variables, you're bound to come across some really crazy circular plots. It's not clear what these curiosities are supposed to reveal, but they sure look cool! So that you won't feel left out when everyone else is showing off their crazy circular plots, I've provided a program (crazyplot.c) to assist you in creating as many as you could possibly need. You will find it in folder examples\crazyplot. Some examples follow. The functions that produce these are shown below the plot.

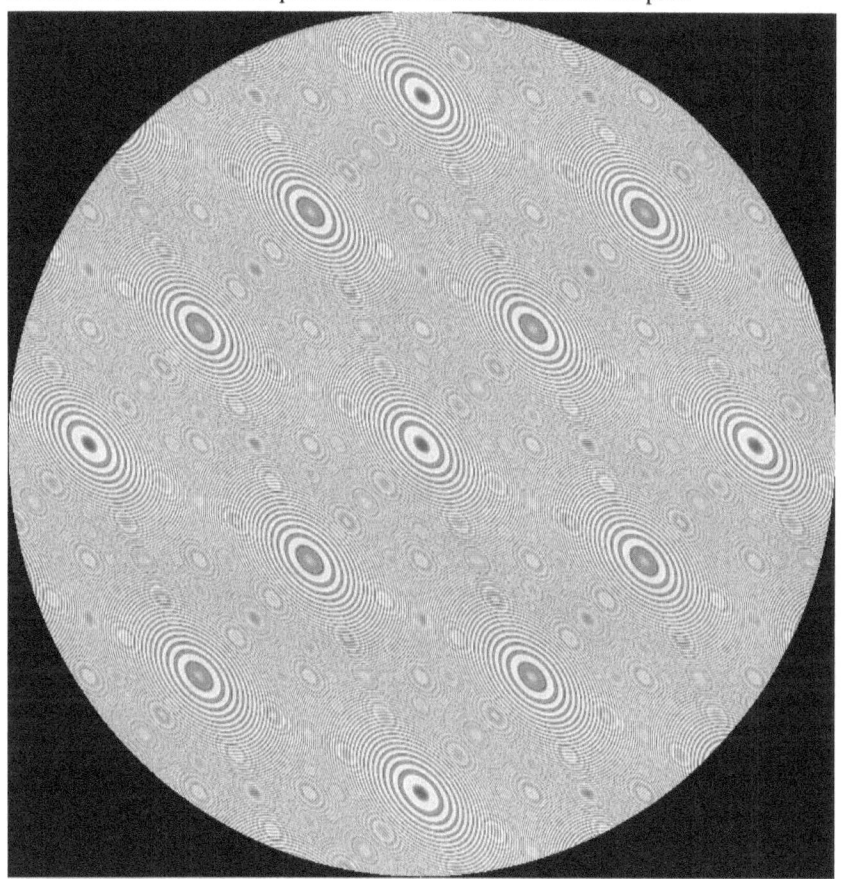

(x*x+x*y+y*y)/2

x/exp(-abs(y)/128)+y/exp(-abs(x)/128)

(x*x*x-y*y*y)/65536

(x*x+4*x*y+y*y)/2048

sqrt(fabs(x*x+y*y)/2)

sqrt(fabs(x*x-y*y)/3)

64*log(fabs((x+x*y+y)))

(x+y)*cos((x+y)/256)

## also by D. James Benton

*3D Articulation: Using OpenGL*, ISBN-9798596362480, Amazon, 2021 (book 3 in the 3D series).

*3D Models in Motion Using OpenGL*, ISBN-9798652987701, Amazon, 2020 (book 2 in the 3D series.

*3D Rendering in Windows: How to display three-dimensional objects in Windows with and without OpenGL*, ISBN-9781520339610, Amazon, 2016 (book 1 in the 3D series).

*A Synergy of Short Stories: The whole may be greater than the sum of the parts*, ISBN-9781520340319, Amazon, 2016.

*Azeotropes: Behavior and Application*, ISBN-9798609748997, Amazon, 2020.

*bat-Elohim: Book 3 in the Little Star Trilogy*, ISBN-9781686148682, Amazon, 2019.

*Boilers: Performance and Testing*, ISBN: 9798789062517, Amazon 2021.

*Combined 3D Rendering Series: 3D Rendering in Windows®, 3D Models in Motion, and 3D Articulation*, ISBN-9798484417032, Amazon, 2021.

*Compression & Encryption: Algorithms & Software*, ISBN-9781081008826, Amazon, 2019.

*Computational Fluid Dynamics: an Overview of Methods*, ISBN-9781672393775, Amazon, 2019.

*Computer Simulation of Power Systems: Programming Strategies and Practical Examples*, ISBN-9781696218184, Amazon, 2019.

*Contaminant Transport: A Numerical Approach*, ISBN-9798461733216, Amazon, 2021.

*CPUnleashed! Tapping Processor Speed*, ISBN-9798421420361, Amazon, 2022.

*Curve-Fitting: The Science and Art of Approximation*, ISBN-9781520339542, Amazon, 2016.

*Death by Tie: It was the best of ties. It was the worst of ties. It's what got him killed.*, ISBN-9798398745931, Amazon, 2023.

*Differential Equations: Numerical Methods for Solving*, ISBN-9781983004162, Amazon, 2018.

*Equations of State: A Graphical Comparison*, ISBN-9798843139520, Amazon, 2022.

*Evaporative Cooling: The Science of Beating the Heat*, ISBN-9781520913346, Amazon, 2017.

*Forecasting: Extrapolation and Projection*, ISBN-9798394019494, Amazon 2023.

*Heat Engines: Thermodynamics, Cycles, & Performance Curves*, ISBN-9798486886836, Amazon, 2021.

*Heat Exchangers: Performance Prediction & Evaluation*, ISBN-9781973589327, Amazon, 2017.

*Heat Recovery Steam Generators: Thermal Design and Testing*, ISBN-9781691029365, Amazon, 2019.

*Heat Transfer: Heat Exchangers, Heat Recovery Steam Generators, & Cooling Towers*, ISBN-9798487417831, Amazon, 2021.
*Heat Transfer Examples: Practical Problems Solved*, ISBN-9798390610763, Amazon, 2023.
*The Kick-Start Murders: Visualize revenge*, ISBN-9798759083375, Amazon, 2021.
*Jamie2: Innocence is easily lost and cannot be restored*, ISBN-9781520339375, Amazon, 2016-18.
*Kyle Cooper Mysteries: Kick Start, Monte Carlo, and Waterfront Murders*, ISBN-9798829365943, Amazon, 2022.
*The Last Seraph: Sequel to Little Star*, ISBN-9781726802253, Amazon, 2018.
*Little Star: God doesn't do things the way we expect Him to. He's better than that!* ISBN-9781520338903, Amazon, 2015-17.
*Living Math: Seeing mathematics in every day life (and appreciating it more too)*, ISBN-9781520336992, Amazon, 2016.
*Lost Cause: If only history could be changed...*, ISBN-9781521173770, Amazon, 2017.
*Mass Transfer: Diffusion & Convection*, ISBN-9798702403106, Amazon, 2021.
*Mill Town Destiny: The Hand of Providence brought them together to rescue the mill, the town, and each other*, ISBN-9781520864679, Amazon, 2017.
*Monte Carlo Murders: Who Killed Who and Why*, ISBN-9798829341848, Amazon, 2022.
*Monte Carlo Simulation: The Art of Random Process Characterization*, ISBN-9781980577874, Amazon, 2018.
*Nonlinear Equations: Numerical Methods for Solving*, ISBN-9781717767318, Amazon, 2018.
*Numerical Calculus: Differentiation and Integration*, ISBN-9781980680901, Amazon, 2018.
*Numerical Methods: Nonlinear Equations, Numerical Calculus, & Differential Equations*, ISBN-9798486246845, Amazon, 2021.
*Orthogonal Functions: The Many Uses of,* ISBN-9781719876162, Amazon, 2018.
*Overwhelming Evidence: A Pilgrimage*, ISBN-9798515642211, Amazon, 2021.
*Particle Tracking: Computational Strategies and Diverse Examples*, ISBN-9781692512651, Amazon, 2019.
*Plumes: Delineation & Transport*, ISBN-9781702292771, Amazon, 2019.
*Power Plant Performance Curves: for Testing and Dispatch*, ISBN-9798640192698, Amazon, 2020.
*Practical Linear Algebra: Principles & Software*, ISBN-9798860910584, Amazon, 2023.
*Props, Fans, & Pumps: Design & Performance*, ISBN-9798645391195, Amazon, 2020.
*Remediation: Contaminant Transport, Particle Tracking, & Plumes*, ISBN-9798485651190, Amazon, 2021.

*ROFL: Rolling on the Floor Laughing*, ISBN-9781973300007, Amazon, 2017.
*Seminole Rain: You don't choose destiny. It chooses you*, ISBN-9798668502196, Amazon, 2020.
*Septillionth: 1 in $10^{24}$*, ISBN-9798410762472, Amazon, 2022.
*Software Development: Targeted Applications*, ISBN-9798850653989, Amazon, 2023.
*Software Recipes: Proven Tools*, ISBN-9798815229556, Amazon, 2022.
*Steam 2020: to 150 GPa and 6000 K*, ISBN-9798634643830, Amazon, 2020.
*Thermochemical Reactions: Numerical Solutions*, ISBN-9781073417872, Amazon, 2019.
*Thermodynamic and Transport Properties of Fluids*, ISBN-9781092120845, Amazon, 2019.
*Thermodynamic Cycles: Effective Modeling Strategies for Software Development*, ISBN-9781070934372, Amazon, 2019.
*Thermodynamics - Theory & Practice: The science of energy and power*, ISBN-9781520339795, Amazon, 2016.
*Version-Independent Programming: Code Development Guidelines for the Windows® Operating System*, ISBN-9781520339146, Amazon, 2016.
*The Waterfront Murders: As you sow, so shall you reap*, ISBN-9798611314500, Amazon, 2020.
*Weather Data: Where To Get It and How To Process It*, ISBN-9798868037894, Amazon, 2023.

www.ingramcontent.com/pod-product-compliance
Lightning Source LLC
Chambersburg PA
CBHW030723220526
45463CB00005B/2156